MICRO SPACE CRAFT

MICRO SPACE CRAFT

Rick Fleeter

The Edge City Press
Reston, Virginia

Trademarks: The following trademarks appear throughout this book: AST, Apple, Mac, Microsoft, Macintosh, IBM, Mustang, Buick, Dixie, Alfred E. Neumann, Star Wars, *Vogue*, 7-Up, *Playboy*, Cray, Miss Manners, Four Seasons, Elmer's, Domino's, Coca-Cola, Pentium, Hitachi, Peugeot, Tylenol, Bristol Aerospace, Disneyland, Metallica, MacDonnell Douglas, Eveready, Star Trek, Dodge, Bon Jovi, MTV, Lucky Strike, Carmen San Diego, Monopoly, Scrabble, Quicken, TRW, Boeing, Ball, Motorola, Moog, Eniac. All brand names and product names used in this book are trademarks, registered trademarks, or trade names of their respective holders. Edge City Press is not associated with any product or vendor mentioned in this book.

Limit of Liability/Disclaimer of Warranty: The author and publisher have used their best efforts to insure accuracy and completeness in preparing this book. Edge City Press and the author make no representation or warranties with respect to the accuracy or completeness of the contents of this book and specifically disclaim any implied warranties of merchantability or fitness for any particular purpose and shall in no event be liable for any loss of profit or any other commercial damage, including but not limited to special, incidental, consequential, or other damages. Any slights against people or organizations are unintentional.

Micro Space Craft. Copyright ® 1995 by Rick Fleeter.

Printed and bound in the United States of America. All rights reserved. No part of this book may be reproduced in any form or by any electronic or mechanical means including information storage and retrieval systems without permission in writing from the publisher, except by a reviewer, who may quote brief passages in a review. Published by Edge City Press, 10912 Harpers Sq. Ct., Reston, VA 22091. First Edition.

Page layout and editing by T. L. and F. S. Ponick of Edge City Press.

Cover and illustrations by Heidi Given.

99 98 97 96 95   5 4 3 2 1

ISBN: 0-9648242-0-5

For information on purchasing this book, contact Edge City Press at 703/620-6650.

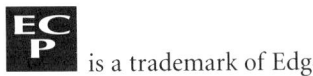 is a trademark of Edge City Press.

About the Author

Dr. Rick Fleeter is a founder and President of the small satellite and space transportation company AeroAstro, and the International Small Satellite Organization (ISSO). He has been responsible for development of over 20 miniature satellites ranging from 2.5 to 250 pounds and has been writing and publishing the *New Space* (previously ISSO) newsletter bimonthly since 1987.

At TRW and DSI, Rick pioneered early defense applications of miniature satellites and developed new satellite propulsion systems and chemical lasers. Rick received a commendation for contributions to the successful rescue of TRW's TDRS-1 communications satellite. He was Senior Scientist at the Jet Propulsion Laboratory of Caltech and since 1979, contributed to AMSAT, the world's most experienced small satellite organization.

Rick has authored numerous papers on thermodynamics, propulsion, and small satellite design. He wrote the small satellite systems engineering chapters of the Air Force Academy textbook, Spacecraft Mission Analysis and Design, and an upcoming Air Force text devoted to small spacecraft. Rick received engineering PhD and AB degrees from Brown University and the MSc in Aerospace Engineering from Stanford. He taught undergraduate and graduate courses in thermodynamics at UCLA and Cal State Long Beach.

Rick writes, swims, and bike commutes in Reston, Virginia, where he lives with his wife, Nancy, Financial Controller of the John F. Kennedy Center for the Performing Arts in Washington, DC.

Here's What People are Saying about Micro Space Craft:

"One Helluva book. Makes absolute sense — required reading for anyone needing an understanding of space. Gives the lay or professional person a great sense of the space world. It grabs you and holds you — read two pages, read just a half page, and you'll read the whole book. The only book on satellites you can't put down."

— *Professor Rudy Panholzer, U. S. Navy Post Graduate School*

"This book could turn out to be the 'security blanket' of small satellite builders — something to grab and hold tight when things get rough. The New World Order of Technology is that the cutting-edge technology of this morning can be mastered by a garage-sized operation after lunch. This book shows how this works in the area of building satellites."

— *Sven Grahn, Swedish Space Corporation*

"A great slant on the technical. I'm relearning a lot of science as I read through it. I'd like my daughter to start reading it also as a text. This could be the 'Zen and the Art of Motorcycle Maintenance' of our generation. It's great."

— *Luther Briggs, U. S. Air Force Space Warfare Command*

"The most fun I've ever had reading about satellites, rockets, and space. Required reading for anyone interested in space — high school students to college professors, scientists, engineers, and high-tech managers."

— *Professor Bob Twiggs, Dept. of Aeronautics and Astronautics, Stanford University*

Introduction:

The way we (like to think we) were

Sizzling in the hot Virginia summer sun and humidity, I try to imagine January. Let's see: snow piled up over the gas grill; icicles hanging from the gutters; driveway a mixture of ice, snow, and largely ineffective sand; the non FWD (TWD?) car cowers in the garage; ducks face windward standing on sheets of ice on the pond out back. All the images are there, but without the bite. The twain, even in imagination, don't fully meet.

Can we remember when space was mission, rather than budget, driven? In the late '80s Air Force generals pronounced small, low cost spacecraft "worthless" and a "distraction." Today, excepting some special cases, if a program can't lay a claim to smallness, low cost, and rapid schedule, it is either dying or already dead.

Aerospace's fall from grace with the sea of humanity

Why? Shrinking budgets is the obvious and partially true answer. Aerospace's springtime of raw attractiveness for its own sake is over. Society has been there, done that, and won't spend large sums on an infatuation with the siren song of space. The dollars are smaller now, and relevance to more people means moving projects away from the exclusive purview of a few industrial and scientific giants. Relevance in part forces lower cost, more rapid schedules, and a more diverse community of developers including students and young professionals — people who don't have decades in one job to nurse along, stone by stone, construction of a modern pyramid. Immediacy and relevance are attributes of low-cost programs of more limited scope.

The world has changed in other ways than just economic. The rate of change, in science and particularly in technology, is increasing. Classical aerospace programs require decades from concept to flight. In that time they are typically obsoleted by several generations and often uninteresting beyond an initial "turn it on and see if it works." It has become more important to fly a new technology immediately, learn and progress forward, than to wait long years to create and fly the most capable possible facility.

Savoir Faire - Can't leave home without it

These are societal imperatives. But in our technology-based society, change happens only when there is a confluence of our desires and our technical abilities. What created mass markets for answering and fax machines in the '80s? The desire to stay in touch coupled with the technical ability to improve our connectedness. In the '90s the cellular phone is booming because our ability to provide mobile telephone service complements our increasing mobility. Small satellite technology is now so important both because we want it (all those societal factors) and we can do it — a highly capable spacecraft which a few years ago weighed tons and was built by thousands can now be built, thanks to revolutionary progress in electronics and the ingenuity of small satellite designers, by 10 people on a tabletop.

No Fear, No Limits, No Grades

Except for occasional (ok, frequent) philosophical interludes, this book is an introduction to the core technologies behind small satellites. It is not a textbook, it does not profess great rigor and it is far from exhaustive — where necessary I may have shortchanged Fourier in favor of Fun and taken a few scenic but definitely off-road shortcuts to get from Anabru to Zurbranchburg without straining any attention spans. But it covers the entire field of low-cost space activity in a way that is meant to be accessible — as are the satellites themselves — to a wide audience. This is satellite engineering for the rest of us — not just for people who've spent 5 or 10 years studying the subject in college and in engineering jobs.

We have an 800 number

Some of the topics covered have been serialized in The ISSO (International Small Satellite Organization) Newsletter, recently renamed *New Space*. All of them have been updated and improved for this book (for more information on *New Space* and its Web site and newsletter, call 1-800-636-4776). The book chapters stand alone, and the bitesize morsels of education they contain include no prerequisites and no grades — pick it up wherever you'd like and browse.

I wasn't born this way

Thanks to Professor Dan DeBra, whose classes at Stanford were a delight and an inspiration, and who encouraged the articles that led to this book. The only thing Dan showed me I ever found incomprehensible is his stamina on a bicycle. Professor J. Kestin taught me the discipline to write and to continue to learn. I am grateful to have been allocated a small part of his time while he was alive among us, time which could have been spent on more serious and more talented students. Another inveterate biker, Richard Warner, encouraged and co-founded our company, AeroAstro. Richard suggested many of the book's topics, created several illustrations, and co-authored with me several sections. This plus his predictably unpredictable advice has been invaluable. I thank Josh Cohen who composes, out of the ASCII mess I create and e-mail to him from long distance and noisy phone lines, readable, interesting, and attractive articles. Professor Rudy Panholzer of the Naval Postgraduate School first suggested my writing this book — a suggestion I found easily dismissable at the time, but which ultimately seeded the entire effort. Despite my decidedly anti-social tendencies, two women have taken an interest in my life and made a difference in this book. My mother, who since 1967 has insisted my writing is great, even when I know it's lousy, who also claimed my piano playing was great even when it was truly unbearable, and who has kept AeroAstro together by providing us an unending supply of the world's greatest bread, aka Hannah's Bananas. Finally thanks to my wife (albeit at low-duty cycle owing to my travel, writing, and work schedules) Nancy, who has learned to tolerate loss of a husband in exchange for an icon of same. One which mainly sits at the kitchen table, eats watermelon, stares at a computer screen, and taps at its keyboard.

—Rick Fleeter

Table of Contents

INTRODUCTION..vii

CHAPTER 1: WHY ARE WE HERE?......................1

CHAPTER 2: PROPULSION—OR— HOW TO GET THERE?........................7
   Rocket Travel Advice: Pack Light........................11
   More Rocket Travel Advice: Carry a BIG Slingshot..........12
   The One and Only Equation..............................14
   Thrills, Chills, and Spills with the Rocket Equation..........15
   Next: A Bit More Detail on Isp..........................17
   Score: Your Lot in Life.................................18
   Anatomy of a Rocket...................................20
   What to Feed a Rocket: Propellants.....................22
   Life Cycle of the Thermodynamicist......................24
   Your Mission: Choosing a Rocket Technology..............33
   Earth to Orbit..33
   What Really Happens When People Build Rockets..........37
   Upper Stage and High Performance Rockets...............38
   Tiny Rockets...39
   The Final Frontier: Infinite Isp..........................41
   Propulsion for Small, Low Cost Missions..................42
   See You in the Hot Tub for Some Gluhweine!..............43

Micro Space Craft

**CHAPTER 3: ORBIT MECHANICS—OR—
    WHAT KEEPS THESE THINGS UP, ANYWAY?** 45
    T-Off Time. 46
    Orbit Definition Good Enough for the Rest of Us 47
    How Mother Nature is Cruel to Eight-Year-Olds 48
    Orbit Altitude. 50
    So What's Everybody Doing Up Here? 52

CHAPTER 4: ORBIT MECHANICS II: The Movie 53
    Geosynchronous Orbits (GEO). 54

CHAPTER 5: YOU SEND ME: Orbit Mechanics III 61

**CHAPTER 6: MAGNETIC ATTRACTIONS
    (with Richard Warner)** 67
    Introduction to Alchemy, Magnetism, and Cold Fusion. 67
    What Rocks Tell Us 70
    Magnetic Many Uses Game 71
    Assume a Can Opener 72
    Satellite Compass: the Magnetometer 73
    Navigating by Magnetic Compass 75
    Can You Say "Torque Coils"? 77
    Magnets Chasing Their Own Tails. 78
    The Force: It Comes in Colors. 79
    For More Info. 81

**CHAPTER 7: EVERYTHING YOU ALWAYS WANTED TO
    KNOW ABOUT RADIO, Part I:
    Shatter the Myth of the Digital Miracle?** 83

**CHAPTER 8: EVERYTHING YOU ALWAYS WANTED TO
    KNOW ABOUT RADIO, Part II:
    Faster Than a Speeding Bullet** 91

Table of Contents

CHAPTER 9: EVERYTHING YOU ALWAYS WANTED TO KNOW ABOUT RADIO, Part III: What's Up, Doc? 99

CHAPTER 10: THERMAL DYNAMICS: Tough Talk About Temperature (A Short, Virtually Painless, and Occasionally Philosophical Look at Spacecraft Thermostatics and Thermodynamics) 109

CHAPTER 11: YOU GOT AN ATTITUDE, BUDDY? (A Primer on Small Satellite Stability and Control) .. 123
    Active Control 127
    Spin Stabilization 132
    Conclusions 137

CHAPTER 12: MEMORY SYSTEMS FOR SPACECRAFT—OR—MEMORY—WHAT IS IT GOOD FOR? (With Richard Warner) 139
    History 141
    What's Available in Satellite Memory Devices? 142
    Anatomy of a Solid State Memory System 144
    Semiconductor Memories 144
    DRAM Storage 144
    EPROM Storage 145
    SRAM Technology 146
    Radiation Effects 146
    Error Detection and Correction 147
    Transfer Rate 149
    Interfaces 149
    Power Consumption 150
    Testing, Quality Assurance, and Reliability 150
    Additional Processing Tasks 151
    Vendors 151
    Case Study I: ALEXIX/MOXE/EUVITA 151

CHAPTER 12: MEMORY SYSTEMS FOR SPACECRAFT (cont'd.)
    Case Study II: HETE/QUEN 153
    The Future? 153

CHAPTER 13: MECHANISMS:
The Nuts and Bolts of Small Satellites. 155
    Moving Parts 155
    Shall We Confront Our Fears? 157
    What's Out There? 163
    Testing 165
    A World of Mechanisms 166
    You Turn Me On: Rock & Roll and Explosive Bolts 167

CHAPTER 14: BATTERIES NOT INCLUDED 173
    When You're Away 174
    Let the Sun Shine 174
    Have You Ever Had to Make Up Your Mind? 175
    How Do They Stack Up? 176
    You Say Tomato, and I Say Tomahto 177
    Qualify, Qualify, Qualify 180
    This Year's Model 180
    Qualification II - The Sequel 181
    Solar Arrays for Small Spacecraft. 182
    Requisite Power Level and Mission Lifetime 182
    Operating Temperatures 183
    Spacecraft Configuration 183
    Array Configuration 183
    Energy Conversion Efficiency 184
    Radiation Degradation 185
    Illumination and Orientation 185

CHAPTER 15: BRING 'EM UP CLEAN 187

CHAPTER 16: SATELLITE CLUSTERS 197
    What Are These Satellites Doing to Keep Busy Up There? 200

Table of Contents

CHAPTER 17: WHERE TO LOOK FOR HISTORICAL UNDERPINNINGS, TERM DEFINITIONS, AND REVOLUTIONALY ZEAL TURNED UP TO 11......205
First, Some Definitions of Small Satellites..................205
Small Spacecraft Time Capsule: The Way (We Think) We Were 209

CHAPTER 18: SPACE HISTORY AND A POSSIBLE FUTURE......................219

INDEX 223

Chapter 1
Why Are We Here?

PARENTS like to think of their children as more handsome, more successful, and happier than the parents themselves are, but nowadays especially the focus on offspring is not so much achieving Lake Wobegone "above average" status, but rather consummate genius. The misconception that satellite engineering is a fantastically complex undertaking has really been a boon to my own parents' wish fulfillment, and probably many others.

Filial piety being a concept reserved not solely for the Orient, it is easy to ignore how useful some rather simple, inert satellites actually are. The moon comes to mind in this regard. It has no digital electronic systems, no radios, no photovoltaics, no batteries. The moon sports not a single moving part ("active component" in today's jargon).

Despite having not a line of software to its credit, the moon is more useful than most of us acknowledge. It provides an efficient, even aesthetic, nightlight. It is a passive reflector used to relay radio signals between distant places on earth. Landing on it provided a major challenge to our own space technologies, not to mention returning from it. The moon's gravity has been used to assist interplanetary spacecraft on their trajectories and to test Einstein's gravitational theories. The moon also creates the earth's oceans' tides, themselves pretty handy not just for creating beaches, inlets, and breeding territories

The moon: an ecologically friendly nightlight

for animals, but also for producing electricity and for harvesting clams. The moon is incredibly reliable, with an MTBF (mean time between failures) measured in billions of years.

In the future we could be lucky enough to find ice on the moon, or maybe minerals containing useful substances like oxygen to use in establishing a lunar outpost. It can be argued that building a space station is redundant, because the moon provides a ready base for operating in space.

The moon is such a clearly good idea that it is interesting to wonder what the world would be like if the moon weren't there and NASA were to propose building it. The environmental impact studies alone would cost billions. It is doubtful that building a moon would ever be approved. Fishing interests would decry the moon, claiming its tides would destroy ocean navigation as we know it. Astronomers would point out that a significant percentage of viewing opportunities would be totally spoiled. I suppose artists would decry the loss of night's velvet cover of darkness, unaware of the charm a large yellow moon can add, hanging huge over the horizon, to a warm summer evening, or the mystic chill it gives us when it is high overhead of a midwinter night's snowscape. The moon speaks to us without radios and in its own language, a language people, plants, and animals all understand.

Technology changes things, even low-tech stuff like earth's moon, and people don't like change. As a technologist, you have to accept that and work with it, realizing that nobody is going to thank you for delivering change, even change for the better. Know any cities or countries named after scientists and engineers? Einsteinville? Archimedesland? Gausstown? People prefer politicians, despite what you read in the press. Leningrad (gone but not forgotten), The Tom Bradley International Terminal at LAX, Cape Kennedy. Politicians are predictable, and they haven't made anything fundamentally better, or really changed anything about our lives, in hundreds, even thousands of years. At best they cook stuff up with the ingredients society provides them, but they are players on the existing field. They don't create new worlds, or even new fields.

Chapter 1—Why Are We Here?

If it's not too late, you might consider violin making. If you can figure out how to make violins exactly like they did 500 years ago, you'll be rich, famous, and popular beyond your wildest dreams. But if you invent a new piano that is played beautifully, mysteriously, hauntingly, without touch or training, simply by modulation of your own brain waves, the art world would assassinate you for alleged debasement of human culture. Have you ever walked into the office of a CEO of a major corporation, or an Ivy League college dean's office, or the office of a celebrated politician and found a bunch of aircraft radios, gyrocompasses, and air navigation charts festooning the place? Maybe a GPS receiver? Of course not, they're all new and crass. But if you find a yellowed old globe missing most of the detail of the New World, maybe a wooden ship's wheel, and a compass that Columbus could have used, complimented by a worn, useless brass sextant thrown in for good measure, that wouldn't surprise you at all. Change has very little constituency. You have been warned.

The moon has an important lesson for we who build artificial satellites. Complexity and usefulness are not two sides of the same coin. They are often travelling companions on our technological journeys, but a single drop of water says just as much as the powerful river and the massive ocean it flows into. Where complexity and usefulness part ways is often where you get the greatest overall value. Books, paper, bricks, arrowheads, soap. One of my many language instructors (I needed a lot of tutoring) gave me some good advice. She said, as I left the security of her living room about to cross two oceans and apply my faltering language "skills" on a real world full of native speakers:

"Only say what you know how to say, don't try to say more."

If you want to build a satellite, do it. Build what you can build. A junior high school class can build a satellite. That satellite can be observed and tracked in the night sky or heard on a radio for a few days as it orbits overhead. A single college class could build a satellite with a radio repeater, and a group of students working over several years can build a stabilized platform with a pretty capable computer, digital radios, and some scientific instrumentation.

Micro Space Craft

All of these projects are just as much a valid application of satellite engineering as the most complex devices our societies have produced— satellites like Voyager that sent back images of Saturn and Uranus; Pioneer, which transmits to us today from beyond the edge of our solar system; and TDRS, one of the most complex communications terminals ever built in daily operation from 40,000 km (24,000 miles) above earth.

People build things in a fundamentally different way from nature. We're always in a big hurry. To survive in a Darwinian world, our minds have evolved an intense focus on the individual, and the individual lifespan is short. Similarly, we are preprogrammed to be deterministic. So for us, it's not good enough to wait for a lot of trees to grow along a river, for some of them to die and to fall over in a bunch someplace across that river to create a bridge that our descendants can walk across 100 or 1,000 or 100,000 years from now, even though that certainly is a viable way to create a bridge. Plants work that way. People don't. We chop down a bunch of trees, build a truss structure, and in a few months we have a bridge. In the process, we also create the field of civil engineering and a whole technology grows up—rope bridges, truss bridges, suspension bridges. And we fit bridge technology into the other technologies we have synthesized. For example, we make wide concrete bridges that cars drive over at 65 mph.

Same human behavior at work in space. In 50 years of frantic effort we have surpassed the ancient, majestic moon in several respects. We have satellites that point at the earth, even at special spots on the earth, or at particular stars or planets. We have satellites with active radio receivers and transmitters for relaying television, radio, telephone, and data. Our satellites carry advanced optics for astronomy and reconnaissance, radio receivers for spying on people's telephone conversations, electric power supplies using solar energy, nuclear energy, and chemical energy that we use to power refrigerators, lights, heaters, even personal computers in orbit. Like civil engineering, a whole technology has grown up around how to do stuff in space. Whereas our medieval ancestors studied the moon, the planets and the stars, we study calculus, analytical geometry, dynamics

Chapter 1—Why Are We Here?

of rigid bodies, orbital mechanics, automatic control, and thermodynamics.

But as another of my patient mentors (all my mentors were patient, which sort of says it all) once explained, starting from first principles is fine if you don't want to get past first principles. This is therefore not a book on engineering fundamentals underlying spacecraft engineering; in fact, you might wonder if this is actually a book at all! It's more of a collection of short subjects—sort of a satellite fan's Readers' Digest, or an aerospace engineering equivalent of an aboriginal approach to fishing and hunting. A person who learned from some elders passes on to you the ancient ways of catching a salmon, avoiding crocodiles, or building an igloo. There's a lot of science behind salmon, crocs, and igloos, but neither you nor your mentor is concerned about it. There are pilots and there are aerospace engineers; there are musicians and there are cello builders; there are hungry people and there are gourmet chefs. This book is for people with an appetite for space.

Building, launching, and using spacefaring stuff has developed a pretty significant bag of tricks. The tricks are not magic but there is a great danger in treating them. You can die. Of boredom. OK, I have a short attention span, so that danger lurks ominous in my mind every morning. Treatments in this treatise err on the side of the quick read. Blame it on a spoiled child of the second half of the twentieth century known locally as the author. So my apologies if technical sensibilities are trampled. It's a lesser of many evils.

The bag of tricks includes what we know about orbits and getting into orbits; about radio communication; about keeping our orbiting creations at the right temperature; surviving in the space environment; figuring out in advance of launching something whether it stands an ice cube's chance at a Texas Bar-B-Que in August of surviving "up there"; stabilization and pointing things; space computers; data storage and software; remote sensing from space and other satellite applications; electric power production and storage; why we ever invented clean rooms (and why satellite customers always want to see them). A few geopolitical tidbits are also included like radio frequency allocation; why it is fundamentally impossible to get a small satellite program funded adequately even though it's one hundred times cheaper than large satellite programs; and why the real frontier of satellite technology is to figure out how to build satellites in less than 15 months even though it takes 8 years to get the bureaucratic machinery rolling to turn on the contract in the first place. See above disclaimer on technical purity; this is not the Church of Aerospace Technology. We're closer to the dusty ol' sarsaparilla bar back behind the train station to space. Plant your spurs up on the tabletop, pour yourself a cool one and enjoy the ride.

Chapter 2
Propulsion—or— How to Get There

AN ILLUSION OF YOUTH is that there are right answers. For instance, you know there are the right parents to have; you just haven't got 'em. There are the right cars to drive, but your parents don't drive 'em. There are the right places to be from, but they're always in other countries. At some point, not coincidentally about when you have to actually rent a place and finance a car, reality becomes a process of getting by with the achievable, striving for the desirable, and learning to enjoy the process.

Such was definitely the case with the automobile, and particularly the internal combustion engine and its need for a transmission. The people who first tried to replace steam engines with internal combustion (gasoline) engines in cars had a problem. These early devices operated efficiently only over a very narrow range of engine speeds. Some clever engineer whomped up a set of gears to allow the car to travel at different speeds and placed the gears between the engine and the wheels, along with a "clutch" to separate the engine momentarily so the gears could be switched. This temporary patch, meant only to fix up a small problem with primitive internal combustion engines, is still with us 100 years later in the car's evolution. We use the prettier term "transmission," leaving "gears" for bicycles and gearheads, but call it what you will, and even automate it if you want, it's the patch that became the jeans.

Early propulsion system bug fix.

Micro Space Craft

Propulsion for spacecraft has made similar non-progress over its 10,000-year history. The fundamental problem in space is there's nothing to push against. Fish and swimmers push water around to move themselves. We walk by pushing back on the floor and pushing ourselves forward. We climb ladders by stepping down on their rungs. A bicycle or car tire pushes back on the street to push the car forward. It is not terribly inaccurate to say that aircraft push themselves forward by pushing back on the air, and their wings provide lift by pressing down on the air. In space, except for the occasional photon or errant hydrogen molecule, there is nothing to push on.

Luckily for NASA, the Chinese solved this problem a few millennia ago by building a gadget that brings its own propellant along with it. The propellant is burned, and the hot gases it produces are directed backwards behind the vehicle, pushing the vehicle forward. We call this particular gadget a rocket, since the term "rocket scientist" offers significantly more ego-satisfying gazorch than "gadget scientist." This same performance is repeated every time a rocket is launched. Huge amounts of hot gas are thrown backwards behind the vehicle to push it forward, and we all gasp and reverently stare at this sublime invention of the human intellect.

It's worth noting, however, just how obtuse this "solution" really is. Imagine that instead of driving your car with an engine that turns the wheels that push on the street and move the car forward, you applied a rocketry-based technology. You would carry a mass of propellant, like water, in the car, along with an enormous pump to pressurize it and force it at high speed through a nozzle. With a fantastic effort you could get that water jet up to an exit velocity of 1 km per second (0.6 miles per second or about 2,000 miles per hour). To accelerate your car up to freeway speed, you'd have to chuck about 100 kg (220 pounds) of water overboard. To drive from New York to Boston (300 km) would require pumping a good-sized swimming pool full of water out the rear of the car, requiring frequent stops to refill even a rather humongous water tank, the weight of which itself would require a lot of water to get up to speed. Using an engine in the conventional way, we can achieve this trip with just 20 or 30 easily stored pounds of gasoline (about 5 gallons or 20 liters).

Chapter 2—Propulsion

Is this pretty silly? Put yourself in a rowboat. Maybe you weigh 70 kg and the boat weighs another 30. Instead of carrying your built-in muscles and a few kg worth of oars, you bring a slingshot and a pile of rocks, a big pile that weighs almost enough to sink the boat, maybe 1000 kg, a metric ton. Every time you take a 100 g pebble and shoot it out the back at about 100 m per second (200 miles per hour - a very impressive slingshot!) you accelerate the boat less than 0.01 meters per second (less than 0.02 miles per hour). Besides being a danger to maritime navigation shooting all those pebbles out the back, you'd have no room for passengers or freight.

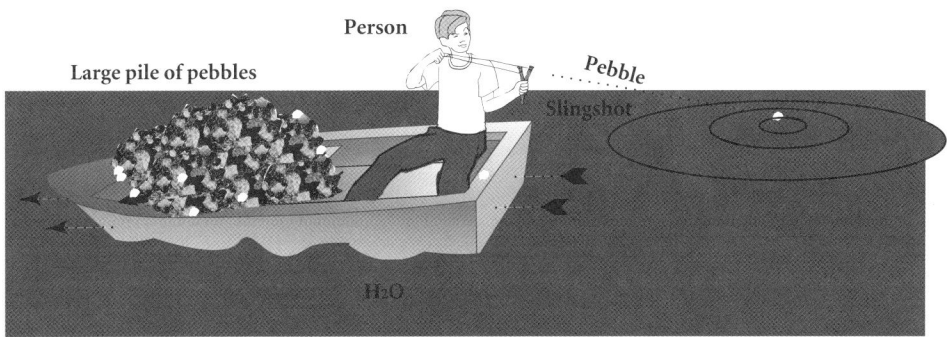

Primitive propulsion system.

What I'm saying is that if rocketry were the only means available to get from A to B, nobody would. Like other inelegant patches we've learned to not just live with but actually respect, like $900 automatic transmissions, telephones that make our voices sound like they are reproduced through string and a Dixie cup, portable computers and CD players with batteries that last 45 minutes, and the housing of our immortal souls in bodies that, with luck, last about 1 part per billion of the lifetime of our surroundings (but nonetheless invariably outlast our savings accounts), we are taught from birth that rockets are really cool devices, worthy of our unquestioning admiration. After all, it doesn't take a rocket scientist to appreciate a pillar of fire and smoke a mile high, even if all that fire and smoke contribute absolutely nothing to moving the rocket on its way except to impress the politicians who allocate money for the next launch.

Micro Space Craft

Like marriage, death, taxes, and college graduation, we got ourselves into this mess with only the best of intentions. No less than Newton himself (the person, not the Personal Digital Assistant) formulated the postulate that to move forward, something has to move backward. Probably even Alfred E. Newman could repeat the immortal words "for every action there is an equal and opposite reaction." All of the examples cited above speak to the fact that you need a medium to push against. Earth, water, air are all very handy. If you have none of these, and space does, which is why we call it space, after all, you have to bring your own.

In fact, Space is not totally empty. It's full of stuff like photons (the quantum mechanical packet of light); a little bit of interstellar gas, but less than one trillionth the density of air at earth's surface; and that universally available ingredient each of us encounters in tremendous supply, wherever we go whether on earth or off it, stupidity. While stupidity drive is not even on the drawing boards yet, the concept of scooping up interstellar gas and throwing it out the back end of a rocket has already been thought of. It works, but. There is so little material out there, that you have to be going sort of WARP-6 (6 times the speed of light) before you bump into enough interstellar matter to make a sensible rocket, leaving us with the gritty little problem of getting from 0 to WARP-6, a speed that Einstein and others have already revealed to us is not attainable due to relativistic considerations. Reality, as they say, bites.

If there is one reason why we don't have 2001, Star Wars, or any other futuristic vision of space activity extant on the particular sheet of the Riemann surface we think of as reality, it is propulsion. Shooting a rocket into orbit is an inaccurate phrase. A rocket lugs itself into orbit as laboriously as we would in a rowboat with a pile of rocks and a slingshot and gives up 99% of its mass by throwing it overboard. Pieces of the rocket itself are even thrown overboard to lighten what remains. I think a precedent for this technology was in the movie "Airplane" where they tossed a bunch of baggage overboard to lighten up the stricken aircraft. If your car were a rocket, it would throw so much of itself away in the process of getting to orbit that only the radio and maybe the steering wheel would actually get there. The rest, including you, would either be converted

Chapter 2—Propulsion

into hot gas, like the rocks in the rowboat, or get dumped to lighten up the "payload" that the rocks were accelerating.

Rocket Travel Advice: Pack Light

Two and only two key factors are under our control in making a rocket as efficient as possible. One is what I have focused on so far, maximizing the fraction of the rocket that is propellant. The more mass you throw backwards, the faster you can move forwards. Rockets typically put less than 1% of their mass in orbit. For very distant missions like going to Mars, the so-called "payload fraction" or "mass fraction" can be less than 0.1%, which is one pound of payload for every 1000 pounds of rocket. An equivalent is building a big four-door Buick whose payload would be maxed out by one skinny Chihuahua or a six pack of Miller. Maybe you remember just how tiny the Apollo service module was perched atop the gigantic Saturn V.

Rocket vehicle designers focus on mass fraction. Their craft is to make as much of the total vehicle mass into propellant, in other words, to minimize the mass of structure, electronics systems, and other dead, meaning non-propulsive, weight. The fact that we actually throw pieces of the rocket away on the way up speaks to the level of desperation modern engineering must go to. We can't afford to take even an empty fuel tank with us. When it's empty, it gets dumped. Needless to say, the tremendous size of rockets relative to their payload and this proclivity for throwing pieces overboard makes rocket travel more costly than planes, trains, and automobiles, not to mention bicycles, surfboards, and elephant backs. It's also less safe if you live under the climbout path, which is why launch sites tend to be at the edges of oceans or deserts.

Lest anyone jump to obvious conclusions, throwing junk away is not the main reason rockets cost big bucks to deliver relatively small payloads to orbit. The main reason is the need to resort to rocketry as the propulsion approach in the first place. Once you break the surly bonds of earth and the atmosphere and get into carrying your own fuel and the oxidizer to burn it, and also the stuff that you want to push back against, you have bought into a very difficult problem. Tossing a few spare parts into the local ocean instead of bringing them back with you costs nothing compared to the cost of building a vehicle that is 90+% propellant and still maintains any payload capacity at all.

More Rocket Travel Advice: Carry a BIG Slingshot

The second parameter determining rocket performance is the one that rocket engine designers focus on. The point of lobbing mass backward is to push ourselves forward. The faster a given mass is lobbed backwards, the greater the acceleration we get from it. A significant fraction of ancient Egypt's labor was directed toward the development of the pyramids, and a significant fraction of the intellectual energy in the United States during the '50s, '60s and '70s, and even the '80s and '90s, has been focused on creating machines that throw mass backward at ever higher speed.

Unlike other modern-day public works projects, rocket performance has at least one easily quantifiable measure of success, which is how fast we are throwing stuff backwards. We could measure a rocket's efficiency by measuring speed like feet or meters per second. This number could pretty much tell you how efficiently you are using the limited fuel resource carried on board. In fact, engineers divide this number, in meters per second, by the gravitational constant, that is, by the acceleration earth's gravity gives to things when they fall. This number, which you might as well know since it is responsible for everything from your fight with the bathroom scale and other sags and bulges characteristic of the aging body, to that family heirloom of your husband's mother's that you unceremoniously knocked off the kitchen counter on your last visit to Cleveland, to the amount of work it takes to get things like rockets off the ground, is 9.8 meters per second per second or 32 feet per second per second. That is, in every second that something is falling, it accelerates another 9.8 meters per second. At the end of two seconds, it is moving 19.6 meters per second, or about 40 miles per

Chapter 2—Propulsion

hour, ignoring effects like atmospheric drag, which is the main reason that potato chips don't shatter when they hit the kitchen floor, but heirloom vases do.

If you divide an exit velocity measured in meters per second by an acceleration measured in meters per second per second, you end up with the units of seconds. Of course if you divide the exit velocity in feet per second by acceleration in feet per second per second, you also get units of seconds. That's a nice thing about this otherwise questionable exercise in division. For example, an exit velocity of 100 meters per second (m/s) divided by 9.8 meters per second per second (m/s^2) gives you a little over 10 seconds. Rocket people call this number the "specific impulse," a very key term in rocketry, because it measures how much propulsive force we squeeze out of every bit of precious propellant. The higher the specific impulse, written Isp, the faster the precious mass we are ejecting is moving, and hence the farther we can go given a certain fuel load. Isp is the miles per gallon of rocketry. Because rocket vehicles are critically limited by the fuel they can carry with them (Remember the prodigious amounts of water needed to operate a rocket automobile?), Isp is the most important single measure of the quality of a rocket.

What are some typical Isp numbers? Some rockets just have stores of pressurized gas, usually nitrogen or helium, that is expelled through nozzles. This is not too efficient, but it's very safe and can be metered highly precisely for close maneuvering. The propulsion packs integral to space walking astronauts' space suits use this approach, and the Isp achieved is typically in the range of 35 to 65 seconds, corresponding to effective exit velocities of around 500 m/s or 1000 miles per hour. That really doesn't sound so shabby, but it's the bottom rung of the rocketry ladder.

Chemical systems that burn various propellant combinations, including the big rockets you see on the space shuttle, achieve Isp ranging from 250 seconds to over 400 seconds with exit velocities over 8000 miles an hour. The highest performance rockets are electrically driven devices that accelerate charged particles (ions, hence the name ion propulsion or ion drive) in an electric field. They can deliver Isp well over 2000s, which is an exit velocity greater than

40,000 miles per hour! But ion propulsion is limited to very small mass flow rates. While the efficiency of propellant usage is very high, the thrust is very low, typically less than 1/1000th of a pound of thrust—less than the propulsive force of a house fly. Nonetheless, applied over many months in space, the minute forces generated by these highly efficient systems can propel spacecraft through the solar system and beyond.

The One and Only Equation

I try to write without equations. Nothing against equations, but people don't like to read them and I like people to read what I write. But there is one equation so basic to rocketry that it's called "the rocket equation." It is simple, yet it contains the Yin and the Yang of rocketry all in one line. The rocket equation describes the relationship between the need for mass fraction efficiency sought by vehicle designers and propellant usage efficiency, Isp, sought by engine designers. Let's just get the equation out there, then we'll talk about it:

$$\Delta V = g\, Isp\, \ln(Mi / Mf) \qquad \textbf{The Rocket Equation}$$

Looks simple. What is it? ΔV is the change of velocity. You start off stationary, light your rocket engine, burn all the propellant, now you're going 5000 miles per hour. Hence your ΔV (pronounced Delta Vee) is 5000 mph. The acceleration given by the earth is a constant known as g. It is 32.2 feet per sec per sec or 9.8 meters per sec per sec. Isp is the rocket engine designer's holy grail, measured in seconds. Multiplied by g it yields units of speed or velocity as meters per second or feet per second. So you can think of the product g Isp as the effective exit velocity of the rocket plume. The higher, the better.

Mi is the mass of the rocket vehicle with all its propellant, with its structure, and with its payload, ready to go. It's the initial mass. Mf is the mass of the rocket exactly like Mi, but after all the propellant has been burned or expelled. It is the final mass.

Chapter 2—Propulsion

Which leaves us with ln, the natural logarithm. Rather than explain logarithms, I'll just mention a few facts about them. There is a button on a scientific calculator labeled ln. If you type in the number 1, the answer you'll get if you then hit ln is O. This means that if the Mf and Mi are the same, you have by definition no propellant on board, and the ΔV is going to be zero. My dad is a busy guy, and he likes to see how much distance he can squeeze out of a tank of gas before stopping at a service station. He is not the type of guy to warm up to a bunch of equations with Greek letters and subscripted variables, but has tested this special case of the rocket equation many times by attempting to continue to drive with Mi/Mf = 1, I.e., no gas. Luckily, he has a cellular phone and a patient wife without a similar inquisitive proclivity.

Thrills, Chills, and Spills with the Rocket Equation

The ln of 2 is about 0.69. How, you might ask yourself, did you ever get as far as you have in life without realizing this? Tack a rocket engine on the back of an old Mustang with the blown engine already pulled. The Mustang, sans engine and all the other parts you gutted from it, like brakes, is left weighing about 1000 lb. Strap on a tank big enough to hold 1000 lb (about 150 gallons, the size of two big hot water heaters) of hydrogen peroxide, a rocket propellant whose exhaust is "just" torch-hot water vapor and hydrogen. Evil

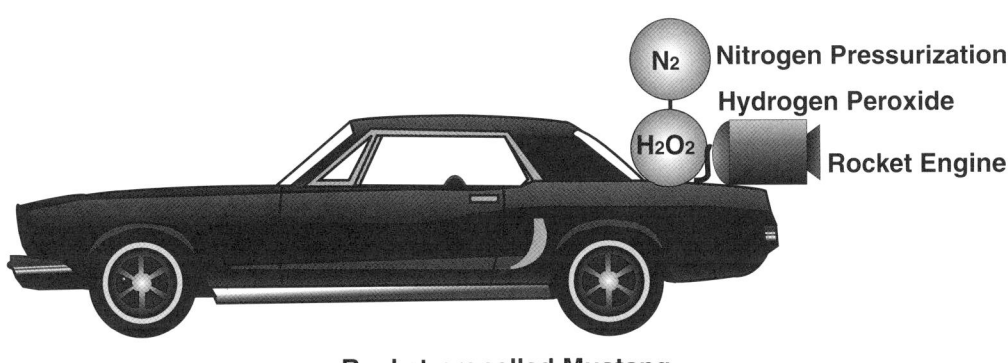

Rocket-propelled Mustang.

Micro Space Craft

Kneivel almost crossed the Grand Canyon on a hydrogen peroxide rocket.

The ratio of Mi divided by Mf is 2. You start off with initial mass of 1000 pounds of Mustang shell and empty tank, plus 1000 pounds of propellant, for a total of 2000 lb. When all the peroxide is exhausted, you've got the 1000 pounds left. Peroxide decomposition is not terrifically energetic, though you may disagree if you spill even a tiny drop on your hands. Typical exhaust velocities are 1500 meters per sec, or about 3000 miles an hour. So the Isp is about 150 seconds.

That's all we need to know. Using the rocket equation, we multiply ln (2), which is about 0.69, times the Isp, 150 s and g, 9.8 m/s/s, which gives a ΔV of 1019 meters per second, or about 2000 miles an hour.

Saying that you should leave this trick to experienced drivers is a bit of an understatement. In fact, at around 250 mph the tires start to melt, the Mustang begins to lift off the track, and air drag gets pretty huge. A lot of your ΔV gets eaten up evaporating rubber, lifting your body and that of your Mustang off the track, and heating the air. You'll be lucky to see 300 mph fleetingly before a messy crash. But you have combined two classic ingredients of drag racing—incredible speed and colossal crackups. If you live, you'll be a hero!

This might sound a bit insane. It is. But it's quite a popular stunt in drag racing, a sport, like many, that redefines the boundaries of sanity. It also illustrates something very important about rockets and the rocket equation. Remember our somewhat clumsy trip to Boston with the water rocket? With just 10 gallons of hydrogen peroxide you can get your '57 Chevy from 0 to 100 mph in less than the time it takes to say "Holy Whiplash, Batman," which is great for space travel. So going to the moon is a matter of applying the right ΔV in the right direction and coasting for a long time. No air or road drag in space—once you impart the ΔV, you coast forever.

But with 10 gallons of gasoline you can shove that Chevy through all the air drag and bearing drag and tire drag and engine drag of a trip from Stanford to Tahoe where you can go up to the top of large,

Chapter 2—Propulsion

snow covered mountains, point your skis straight down the fall line, and personally redefine sanity. What's the difference? There's lots of differences. For one thing, gasoline engines burn air, so they cheat. Rockets carry all their own reactants. The Isp of a gasoline engine can be at least double that of a rocket, if you charge yourself only for the fuel and not the "free" oxidizer. For another thing, its very inefficient, energy-wise, to push yourself forward at 50 mph by throwing rocket propellant out the back at 3000 mph. None of that kinetic energy of the rocket jet, none of its heat and light and smoke producing energy does anything for pushing you frontwards! In a gasoline engine all of that dramatic stuff happens inside the cylinders, where it is captured ultimately as heat that pushes the pistons down and provides motive power. Of course, a properly tuned gasoline engine does not melt the pavement along with any tailgating cars in its wake. The same cannot be assured for even a modest hydrogen peroxide dragster.

Next: A bit more detail on Isp

But first take this test: Put a 1 next to questions you answer TRUE. Put a zero next to questions you answer false:

1) I read the directions when I buy a new Walkman.

2) I actually know how to program phone numbers into my cordless telephone.

3) I've written software for a computer that did something, like played Tic Tac Toe or calculated biorhythms.

4) I sometimes buy small parts (wires, clip leads, etc.) at Radio Shack.

5) I agree with the statement "What's the big deal about programming a VCR?".

Micro Space Craft

6) I have been to an auto junk yard at least once in my life.

7) I have sewn something myself: a sleeping bag, a shirt, blouse, a button…

8) I prefer to eat my own cooking over going to a restaurant.

9) I own more tools than just a pliers and a screwdriver.

10) I can identify the major parts under the hood of my car.

Add up your ones (and zeros if you'd like), then use the grading scale below to know everything necessary to shoehorn yourself into one or two of life's little cubbyholes:

Score: Your Lot in Life

8 - 10 — Face it, you're a hopeless techie at heart. Society is suspicious of you. Avoid politics. Read this section

5 - 7 — You are great to have along on a road trip, but a failure at cocktail parties. Skim this section.

Chapter 2—Propulsion

1 - 4 — You are or will be a highly successful lawyer, accountant, TV personality or businessperson, since you don't get bogged in details. You are highly attractive to the opposite sex, and it's enough to know that this section has some details that engineers worry about. Skip to the next section, Anatomy of a Rocket.

0 — You are destined to lead, possibly as President of the United States or a Vogue model. Definitely skip to the next section: Anatomy . . .

If life were simple (and if turkeys were in fact spherical), the measure Isp would be exactly equivalent to a measure of speed of a rocket's ejected material. But rockets do a little bit more than just accelerate hot material and eject it out the back end of a nozzle. Looking at the illustration below, the inside of a rocket engine is filled with high pressure gases and some other materials. Unlike the more normal pressure vessel shown next to the rocket, an engine has a hole on one side for the exhaust to come speeding out. A normal pressure vessel, like a can of Diet 7-Up or shaving cream, is sealed and static. The pressure forces are equal on all sides of the container, so they cause no net force. The asymmetry of the rocket motor means that the pressure forces are

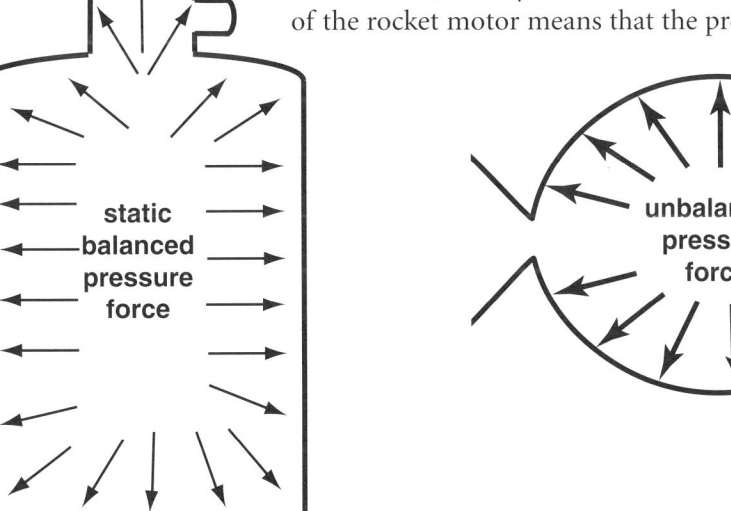

Shaving Cream Propulsion (l) vs. Rocket Propulsion (r). Constant combustion of propellants keeps rocket chamber pressure high despite an open hole on one side.

not balanced, and even if no material were to come out, there would be some propulsive force because of that imbalance.

In practice we measure just two things, the mass flow or consumption rate (how fast we are using up propellant) and the propulsive force. Nobody really cares what the magnitude of the pressure effect and the mass ejection effect are separately. In fact the mass ejection is by far the dominant contributor. Note too that it is impossible to have the asymmetrical pressure distribution without the hole (nozzle throat). Once you have a hole with a high pressure gas inside, mass is definitely going to come out of it, so the two effects are closely linked. We lump the extra oomph gotten from pressure in with the mass ejection derived thrust and treat it as one effect, and from that calculate an apparent exit velocity. The mass consumption divided by the ejection rate, multiplied by the lumped-together exit velocity gives us the propulsive force, and that is Newton's Law. Thus, the propulsive force, including the mass ejection and unbalanced pressure effects together, divided by the mass ejection rate gives an effective or apparent exit velocity. Dividing that exit velocity by the 9.8 m/s gravitational constant gives you the true specific impulse, I_{sp}, of the engine.

Anatomy of a Rocket

The skeleton in the rocket scientist's closet is that rockets are in fact incredibly simple devices. They have only two or three basic parts. They don't have all kinds of spinning wheels and pistons going up and down like airplane gas turbines and automobile engines. They don't have better than one million active switches arranged in complex topographies like microprocessor chips. They don't have zillions of neurons all firing off like the brains of mice and men, or a soup of complex chemical agents like enzymes, proteins, lipoproteins, and complex sugars creating all kinds of growth and motive actions. They have, in essence, just one chamber where fuel and oxidizer are burned and an exit nozzle.

Looking at most rockets is like looking at the engine compartment of your old MG. It's a maze of tubes, valves, nuts, and bolts. The MG is, in fact, complex, albeit needlessly and hopelessly unreliably

Chapter 2—Propulsion

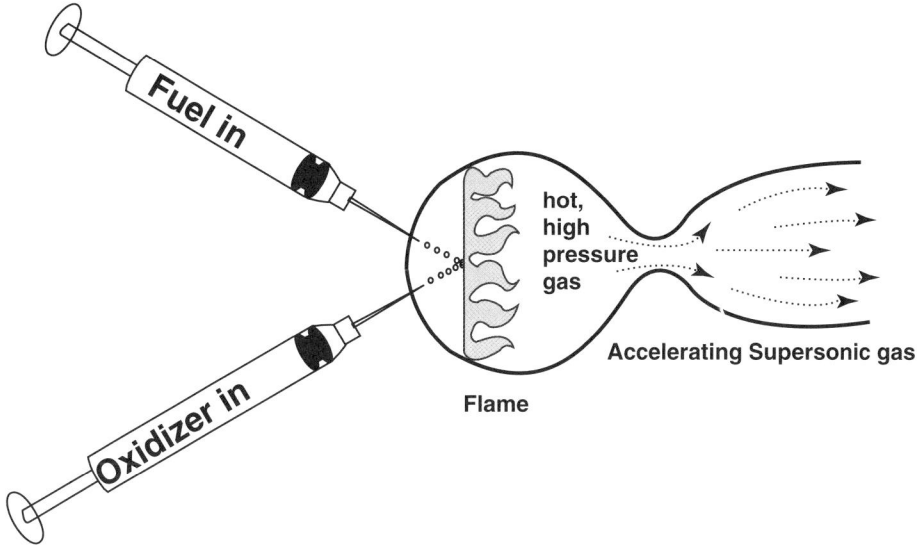

Simplified diagram of a rocket engine. Propellants in, gas out under pressure.

so, but the rocket mainly uses that stuff to cover up its basic simplicity, thereby maintaining the efficacy and clout of the term Rocket Scientist.

The nozzle has a converging, then diverging shape. The propellant gas accelerates to squeeze through the throat of the nozzle where it reaches sonic velocity, that is, the local speed of sound. Gas flowing faster than the speed of sound is further accelerated in the diverging section of the nozzle, sometimes called the bell of the nozzle. Usually the gas inside the engine is under very high pressure, from 10 to 1000 times normal sea level atmospheric pressure. As the gas spreads out in the diverging nozzle, its pressure steadily drops. Rocket engines that operate at sea level, which are the ones that get a vehicle started from the earth, have relatively small bells because they do not want the gas pressure to be lowered below the outside pressure of one atmosphere. Rocket engines that operate in space, where the outside pressure is virtually zero, use much larger bells relative to the size of the engine. By allowing the gas to flow through a larger pressure drop, they achieve about 20% higher exit velocity and higher Isp than their sea-level counterparts.

What to Feed a Rocket: Propellants

Once you get to know them, rockets are as individual as plants, houses, or people. Unlike plants, they don't invade your body with pollen or die when you forget to water them. Unlike houses, they don't have termites or need the grass cut or contain a bunch of relatives with whom you must learn to live in apparent harmony. Unlike people, they don't argue with you about religion or politics or call you during dinner time to sell tickets to the state policemen's dance. In a nutshell, that's why I got into this business. It has afforded me the income to buy a house, surround it with plants, and create the necessity of going to technical conferences to talk about rockets with a bunch of people.

Some plants stand around in your living room at Christmas time and look red and green. Some are farmed in the billions of kilograms to make transfatty acids for people to eat at McDonalds, and some stand around in ancient forests waiting for bulldozers. Similarly with rockets; there are myriad ways of designing and using them. Some rockets are those fiery monsters you see lifting the space shuttle off the ground. Others 1/1000th as big boost satellites from the shuttle to higher orbits. Some once again 1/1000th as powerful are used to gently correct a spacecraft's or an astronaut's position and attitude.

Probably the most common categorization of different rockets is by the propellants they consume, sort of like dividing the dinosaurs into meat eaters, vegetarians, and the omnivorous ones that eat anything. A few of the basic types of rocket engines are mapped out in the table below. All rockets use either a non-reactive propellant like nitrogen gas or some combination of 1, 2, 3, or even more discrete chemicals. Electric propulsion usually uses a single, non-reactive propellant combined with a lot of electric power. Liquid rockets generally carry their propellants as liquids in separate fuel and oxidizer tanks. Solid rockets premix the fuel and oxidizer into a solid crystalline or rubbery material that is molded and bonded into the rocket motor casing itself.

Chapter 2—Propulsion

Categories	Compounds
Inert Gas	• **nitrogen** (simple, safe, I_{sp} around 45s) • **helium** (ditto, I_{sp} around 60s) • **ammonia** (stores more densely but have to separate liquid and gas, I_{sp} 50s)
Liquids	• **hydrazine** (monopropellant, highly toxic I_{sp} 220 s) N_2O_4/mmh (highly toxic and corrosive, but stores permanently at room temperature, I_{sp} low 300s range, expensive) • **LOx kerosene** (LOx must be loaded just before launch, cheap, clean, I_{sp} in low 300s) • **LOx hydrogen** (expensive, hard to make and store, very high I_{sp} - over 400s)
Solids	• I_{sp} upper 200s to 300s
Electric	• **pulsed plasma** • **teflon propellant** (I_{sp} 1000 to 2000s) • **ion thrusters** : Xenon propellant (ditto) arc jets, many propellants (I_{sp} 500s to 1000s)
Hybrids	• **LOx / rubber** (safety comparable with **LOx Kerosene**, now in R&D)

Table of Rocket Propellant Categories and Compounds

It's a world of options, which tells you a few things right there. One is that we really haven't come up with The Solution. For instance, cars. Pretty much they all have four wheels. Yeh, there's vehicles with two wheels, three wheels, 16 wheels, but four wheels has gotten to be pretty dominant. In rockets nothing is truly dominant. For another thing, this chapter could get infinitely long if we're going to cover the field, so we're not and it won't. Trust me on that one.

Micro Space Craft

What with rockets being so ridiculously simple, we thermodynamicists had to figure a few ways to complicate them. You are just not going to get a very big house, afford very many plants, or go to very far away meetings with lots and lots of people at them, if the only knobs available for twiddling are pressure inside a chamber and nozzle geometry. Most of that was worked out in the 18th century, and people are bound to notice that before signing your nth paycheck.

Life Cycle of the Thermodynamicist

Talk about human ingenuity! Hey, we solved this problem years ago. What you do is: You go to public school for 13 years and survive the ire of your peers as you study hard and do well, earning yourself the various titles of nerd, jerk, gearhead, and geek. Then you go to college for four more very similar years of hard work and social rejection, except possibly for the comraderie of a worthless bunch of nerds, jerks, gearheads, and geeks. Then on to four or five more years of graduate school, which is quite a shock after the previous 17 years of rejection for being too brainy. Suddenly everyone around you is telling you that you are probably just too dense to hack it.

After several years of insinuating your inadequacy, they give you the PhD with a smug intimation that it was not so much for your brains or achievement as it was just the most efficient way to rid themselves of you and make room for others more worthy. Then there's maybe a year or two of post-doc-ing, mainly in order to allow you sufficient time to get over the previous five and possibly to discover things called the "out of doors" and something else called "sex." If you don't overdo it too muchly on those two goodies, you emerge into what remains of your youth, or at least young adulthood, handicapped with a world view totally out of synch with what passes for reality.

In society at large, otherwise known as where you are supposed to dwell happily and prosper, you are guaranteed, albeit by people who have never actually tested the hypothesis, that your whole huge bag of new, subtle little thermodynamical tricks will justify and facilitate

Chapter 2—Propulsion

your attainment of the house/plants/meetings/people stuff. All of these attainments eventually consume so much of your time that you really don't have much left for twiddling knobs on rocket engines. Such is life, reduced to one story, albeit short, sweet, and only a little sad, for most thermodynamicists. So before the phone rings, let's have a quick look at those knobs.

One is temperature, which is itself a measure of energy imparted to atoms and molecules. The higher the temperature, the more excited the atomic constituents of the propellant gas get. Hot gases accelerate more rapidly as they travel out the nozzle, converting some of their undirected thermal energy into increased velocity. This effect goes as the square root of the absolute temperature. Typical combustion products can be up to 9 times hotter than room temperature on an absolute scale, so a rocket that burns propellant is about three times more efficient (has 3 times higher Isp) than a rocket using "cold gas," that is, pressurized gas stored at ambient temperature.

Another trick is the design of the propellant molecules themselves. Let me sweep a decade of mind-rupturing study of classical and quantum mechanics, physical chemistry, thermodynamics, statistical mechanics, and mathematics away here and just say that the ideal propellant gas is made up of the lightest, simplest possible molecules (Why didn't they tell me that in the first place?). Why? Molecules have a few choices of what to do with the energy they get from being pressurized and heated. What we want them to do is convert that energy into the highest possible exit velocity. What they also tend to do is convert that energy into things like spinning themselves around, flexing, occasionally breaking their chemical bonds, and vibrating.

Something called equipartition of energy says that for every hour your kid spends doing her homework, she'll spend another hour arguing with you about doing it, an hour on the phone with friends, an hour playing with her food, and an hour in front of the television. Molecules also put about equal amounts of energy into all their alternatives, and why not? From their point of view, one is as useful as another. It is our Western, logical deductive mindset that

Micro Space Craft

tells us that self-rotation, vibration, and even destruction is not productive, but racing out the back of a nozzle at 20 times the speed of sound is highly productive. Scientists are the ultimate imperialists. These internal modes of molecular motion, somewhat akin to a

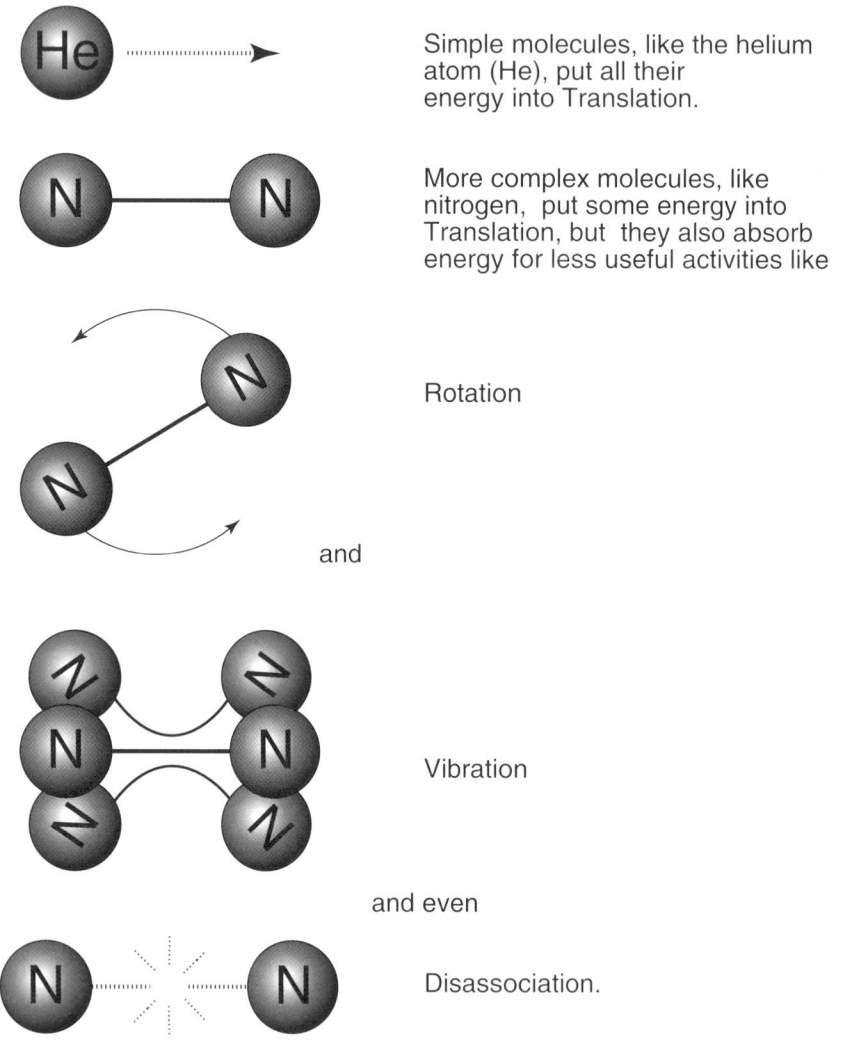

Simple molecules, like the helium atom (He), put all their energy into Translation.

More complex molecules, like nitrogen, put some energy into Translation, but they also absorb energy for less useful activities like

Rotation

and

Vibration

and even

Disassociation.

Molecular Lifestyles.

Chapter 2—Propulsion

proletariat value system with emphasis on God, family, and principle in action, sap energy that otherwise could have been used to accelerate them out the nozzle, grow more rice, increase stock dividends, or whatever…

Playboy claims that hiring only cute women for bunnies is not discrimination, but rather a step in delivering a quality product. Thermodynamicists hide behind the same thin veil of logic, ready to be pierced by some legal or programmatic stiletto. Given the same number of non-productive energy modes, a big fat molecule raised to the same temperature as a trim little molecule goes a lot slower. In fact, the speed is inversely proportional to the square root of the molecular weight at any given temperature. For that reason big molecules, even those with few or no rotation and vibration modes, are very undesirable in chemical propulsion systems. Which is to say, we don't use them except when forced to do so in deference to some external constraint or stimulation.

Nitrogen molecules have a molecular mass of about 28, while hydrogen, which has the same repertoire of vibration and rotation modes, has a molecular mass of 2. This means that hot hydrogen has almost 4 times the Isp of hot nitrogen! Not only that, hydrogen's light atoms and strong chemical bonding make the molecule stiffer and tougher than nitrogen, so it is harder to excite the vibration modes of hydrogen, and much harder to break its bonds, making hydrogen more than 4 times as efficient as nitrogen as a propellant from an Isp point of view. Unfortunately, hydrogen is hard to store. It is only a liquid at temperatures near absolute zero. The liquid is not dense, requiring large, massive, and heavily insulated (that is, even more massive) storage tanks. The gas is very non-dense, so if you don't bother to insulate it, the tanks get really big. Thus although it's an efficient propellant, hydrogen tends to make highly inefficient rocket vehicles. A trade off we'll get into a bit later on . . .

Another problem with big molecules — those with lots of atoms linked together — is that the bonds between the atoms tend to break. Wonderful, subtle, elegant and complex physical reasons explain this, but I'll restrain myself and describe only the two undesirable effects of this tendency of large molecules to come apart, particularly when hot, which they are if we use them in a rocket engine. One is that breaking bonds takes lots of energy, which then isn't available to achieve high exit velocities. The other is that often the new molecules formed when the old one is broken are messy.

For example, molecules we eat tend to be very big, as molecules go. A lot of cooking actually amounts to selective bond breaking. If you leave your toast in too long when the boss phones to complain about the shoddy work you did on the Smith account, you'll return to find that breakfast is a black hunk of inedible stuff. What happened was you overheated the bread and broke a bunch of chemical bonds. A lot of the atoms you wanted to eat escaped as gases, and a bunch of black carbon, bonded tightly by Van Der Waals forces to your best glassware, was left behind. The exact same thing can happen in rocket engines. The leftover carbon not only doesn't contribute to propulsion, but it can clog your engine (sounds like a gasoline ad, doesn't it?).

Unfortunately, chemicals like helium and hydrogen that are light, tightly bonded, and compact tend to be the least dense, hardest-to-package materials. They often are not highly reactive, so it's hard to raise their temperature.

What do you want to look for in shopping for propellant? Something dense with high molecular mass, ergo big molecules and a liquid, or possibly solid, at room temperature. Something stable that won't fall apart, explode things, or eat its container. Preferably something non-toxic that reacts rapidly, but not explosively, with something else dense, stable, and similarly non-toxic. The idea is to create one or more reaction products that are non-dense, meaning they have low molecular mass and a small number of atoms for each molecule; incredibly hot; stable, so they don't react with the materials the rocket itself is made of; and preferably also non-toxic.

Solid exhaust products like carbon and aluminum oxides are a minus. They erode rocket nozzles and decrease performance because they cannot accelerate and expand as gases do when they traverse the nozzle, and because people on the ground don't like them.

Chapter 2—Propulsion

Observe the remarkable parallel between the above list and the dating process. The list of preferred traits is like those ads you see in the personals. Do you think the blonde SWF who is fun loving, leggy, has a great smile, loves kids, the outdoors, and quiet evenings by the fire with someone special, is a professional and a non-smoker and just wants to enjoy life with a SWM of similar persuasion is telling the whole story? Frankly, we haven't found the solution to this problem yet, neither in mate selection nor propellant development, which is why we have such a profusion of candidate rocket propellants, SWFs and SWMs. They each have some attributes and some faults, and we pick for each application one with attributes we need and faults we can live with. You may not have been aware that Dr. Joyce Brothers is not an MD, but a thermodynamicist. You read it here first…

The table up there about fifteen hundred words ago mentions a few of the leading molecules rocket thermochemists selected over the aeons, or at least decades, of rocket research. In terms of pure elegance, my favorite propellant combination is hydrogen and oxygen. They burn with a very hot flame to make water (vapor) as the propellant that comes out the back end. There are absolutely no particles, and no toxic chemicals, either in the reactants or the exhaust. Water has excellent properties as a propulsion molecule. It is very tightly bound and stable and it is light, with a molecular mass of 18. The combination of high flame temperature and small molecular mass gives outstanding Isp - well over 400 seconds.

The major disadvantages of hydrogen and oxygen are that both propellants are cryogenic—not storable for long, and the flame is so hot that it's hard to build engines that stay cool and unmelted! Also, liquid hydrogen is not very dense, making the tankage very large and thus heavy, which costs you some of the great Isp advantage. Also, these propellants don't spontaneously ignite (a property rocket people call hypergolic), meaning they need to be ignited, which is an added complication. An Ariane rocket was lost once because the hydrogen/oxygen upper stage of the rocket failed to ignite. The Space Shuttle Main Engine (SSME) is hydrogen/oxygen, but its clear blue flame is almost invisible in the midst of all the fire and smoke of the solid rockets strapped on to the shuttle.

Another classic propellant is hydrazine, a molecule designed and produced almost exclusively for its value as a rocket propellant. Hydrazine has two nitrogen atoms and four hydrogen atoms, which is the same composition as ammonia, only ammonia has one additional hydrogen atom per nitrogen atom. Hydrazine is a mono-propellant. Most other propellants, like hydrogen and oxygen, are bipropellants; you have to combine fuel and oxidizer to get a flame. Not hydrazine! It decomposes all on its own to form ammonia, nitrogen, and hydrogen, all of which are environmentally benign, and there is no solid component so it's clean burning and efficient. The mean (arithmetic) molecular mass of the exhaust is very low at 16, and the decomposition reaction gives off enough heat to get Isp around 220 seconds, which is pretty respectable without need for carrying two reactants.

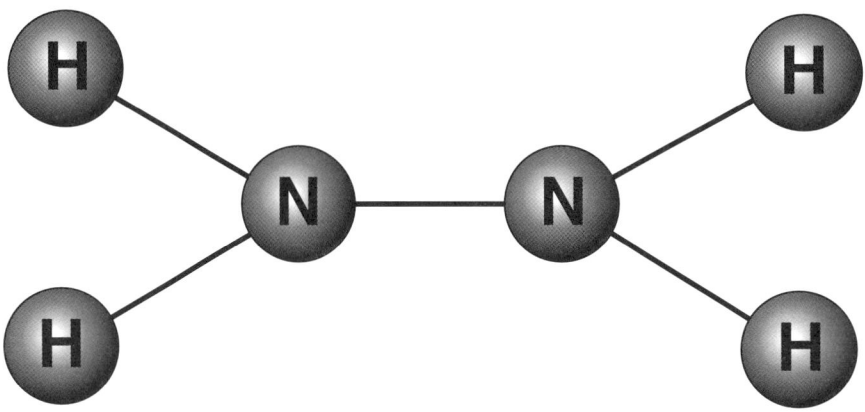

Hydrazine molecule - The classic monopropellant, Hydrazine decomposes to ammonia and nitrogen.

Even though one of the most common propellants in use today, hydrazine has some pretty serious problems. It is immediately poisonous, and, even if you get a dose too low to be lethal now, it is a powerful carcinogen. It is toxic below the concentration threshold to smell it, so if you can smell it, you could be poisoned already. Besides that, it doesn't take much to cause it to decompose, just a trace of ferrous metal like steels and iron, or of carbon, and almost all organic materials have carbon in them. If it decomposes in a

Chapter 2—Propulsion

closed container like a fuel tank, there is a violent explosion. But if the container is 100% free of these contaminants, hydrazine stores forever. This long shelf life is one factor contributing to its use on long duration missions.

While hydrazine is a liquid at room temperature, it freezes at 340F (1•C), so it needs to be kept warm. Because it does not require air or oxygen to burn, extinguishing a hydrazine flame is tricky. Mostly you hose it down with water until the concentration goes down so far that the decomposition stops. All that water is then contaminated and special provisions need to be made to treat the hydrazine before disposal. Because of all these factors, hydrazine handling is a specialty of its own, and the skilled people who do it wear fully protective gear. Setting up a facility for handling satellites with hydrazine takes time, money, and a lot of conversations with people like the EPA. When all that is said and done, it might be no more dangerous than oxygen and hydrogen, which are also easily ignited. At least the decomposition of hydrazine is virtually as clean as that of hydrogen/oxygen, so if you blow yourself up, the neighbors won't be endangered.

Hydrazine can also be combined with oxidizers, particularly nitrogen tetroxide (two nitrogen atoms, four oxygen atoms), as discussed earlier. This combination makes an entirely storable system with good performance (Isp greater than 300s), and it is hypergolic. You get instant ignition when you bring the two chemicals together, but you are handling two volatile, poisonous, corrosive, and potentially carcinogenic liquids, which drives up the cost of the system considerably. This combination of propellants has been used in military missiles, including ICBMs, and many rockets with military heritage, including the entire family of Chinese (Long March) rockets. The propellants are, however, quite expensive, costing over 25 times what oxygen and hydrogen cost for the same total amount of transportation.

Hydrogen peroxide, a water molecule with an extra oxygen atom, totalling two hydrogens, two oxygens, and the subject of our Mustang dragster, is more popular for garage work. The Isp is low (150 seconds or below) but the products, being oxygen and water, are safe. The problems with peroxide, as it's called, are that it too is unstable, particularly at high concentrations. It is normally used diluted in water; contact lens cleaner with 0.5% of peroxide in water is a very vigorous cleaner. For rockets, a minimum is 70% peroxide, and for orbital rockets numbers like 98% are kicked around. Like hydrazine, it decomposes spontaneously, particularly at high concentration. If that happens in a closed container, an explosion is the result. Evil Kneivel used it at 70%. Peroxide can also be used to burn kerosene as a component of an interesting bipropellant. The British Black Arrow got into orbit on that combination, but because you can't store peroxide without suffering continuous decomposition, it is not used much any more.

The workhorse propellant combination is so-called LOx/kerosene, liquid oxygen and kerosene. Not as sexy as LOx/hydrogen, the reagents are cheap and the Isp performance, at around 300 seconds, is very good. The oxygen is not storable, so its primary application is in launch vehicles.

The simplest propellants are inert molecules like nitrogen, helium, and ammonia, which are released through a nozzle—aerosol can rockets. Helium has the highest Isp, but is hard to store because it's not dense, so the tank gets heavy. Ammonia, and also freon and propane, store as much denser liquids, but separating the liquid and gas under zero gravity can be tricky.

Solid propellant rocketry depends mainly on long-chain hydrocarbons as fuel (polymerized kerosene, if you like) and on so-called perchlorates as oxidizers. Often metal particles are added, such as aluminum, or metals are incorporated in the propellants, to raise flame temperature to get higher Isp. The fuel and oxidizer are cast together and need to be ignited, usually electrically, to begin operation. Both the propellants and the reaction products (exhaust) are highly toxic. Creating solids with more benign properties is a subject of current research.

These are just samples of the hundreds of propellants and propellant combinations, but these few constitute a big part of the total field today. As cost becomes more of a driver than military readi-

Chapter 2—Propulsion

ness, we are seeing more attention paid to the oxygen hydrogen and oxygen kerosene propellants, though they are not a solution for on-board propulsion for satellites because they aren't storable. Here the trend may be toward electric propulsion, using propellants that can be teflon rods or inert gases like xenon and argon.

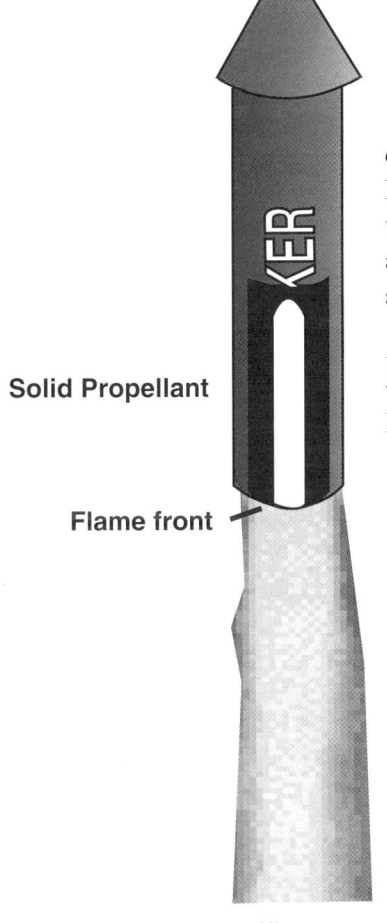

Solid Propellant

Flame front

Hot effluent gases

4th of July Americana derived from ancient China.

Your Mission: Choosing a Rocket Technology

There are three major categories of propulsion applications: launch from earth to orbit, orbit transfer, and orbit/attitude modification. Each application has its own special considerations and optimizes with different propellants and propulsion technologies. Let's shop propulsion systems for these few missions.

Earth to Orbit

Launching into orbit from near earth requires large thrust, larger than the weight of the rocket in order to get it off the ground. That requirement can be a problem for liquid propellant systems, because all the lines carrying fuel and oxidizer must be big enough to handle huge flow rates. You also have to have big, fast pumps to move it and a fuel injector and chamber that can handle it all. All that hardware weighs plenty, making it all the more difficult to get off the ground, requiring yet bigger valves and pumps.

A solution is the solid rocket motor, shown schematically in the figure on the next page adjacent to a liquid rocket system diagram. Solid rocket propellant was used in those 10th century Chinese rockets, and is the stuff amateurs make cardboard rockets fly with (for example, Estes

33

Micro Space Craft

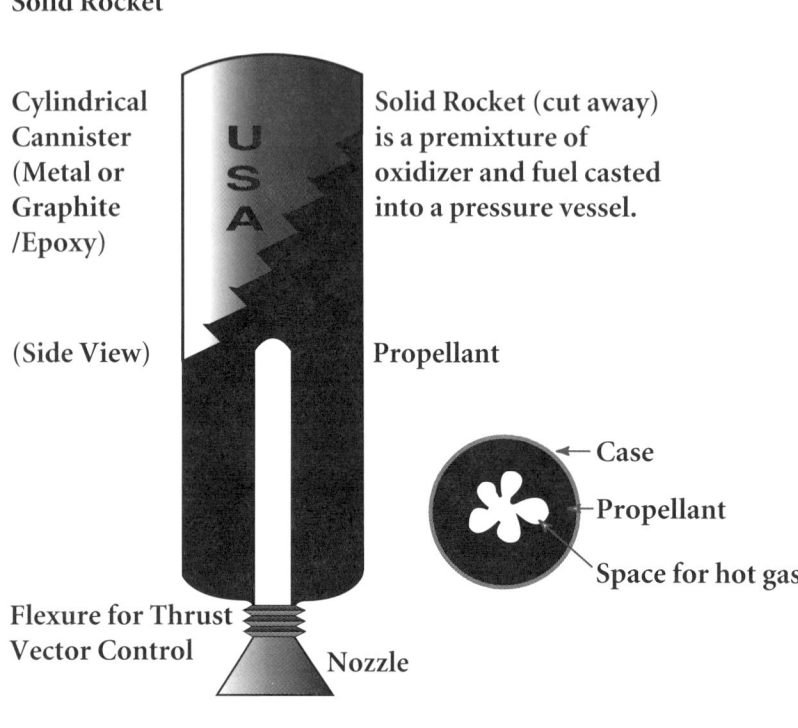

Cutaway diagram of a solid fuel rocket.

Solid Rocket (cut away) is a premixture of oxidizer and fuel casted into a pressure vessel.

rockets), and drives the Pegasus and Scout rockets that have become so important to small satellites. Solid rockets have no pumps, valves or tanks. The oxidizer and the fuel are in the solid phase, and they are pre-mixed into a nearly inert glassy or rubbery substance. Once ignited, however, the premixture burns vigorously and creates a hot exhaust gas that is expelled through a nozzle and produces thrust. Solids don't leak and don't require cold storage, so they are ideal for strategic weapons and other military applications. They can be held ready for launch for years, even decades, and then fired immediately. They are also incapable of leaking, making them quite desirable on board ships and submarines.

For at least the last 40 years, proponents of solid propellant technology have argued these advantages, while liquid propellant rocket companies have squared off to do battle for the contracts to build modern rocketry. This argument is no less likely to be resolved than a final decision on whether Pepsi really out-tastes Coke or All gets collars whiter than Tide. For some applications, solid technology really makes sense. Other times liquids are the only way to go, and

Chapter 2—Propulsion

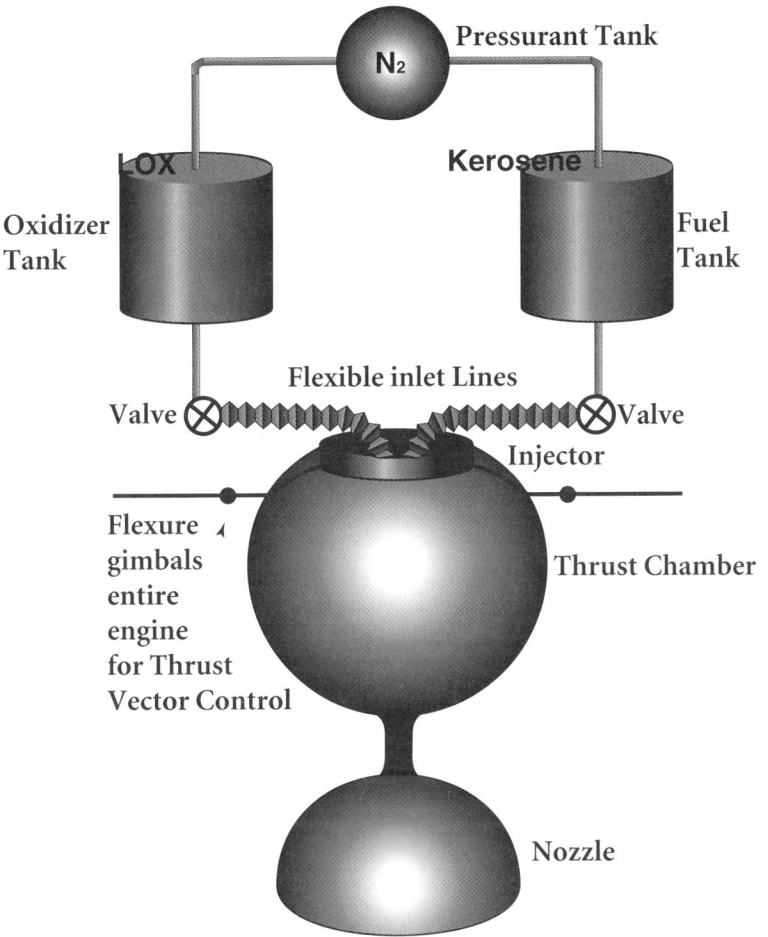

A liquid propulsion system. *Note:* In many rockets, the nitrogen pressurant's job is done by pressurization pumps.

there is a vast territory in which either one could and both have been used successfully.

When shopping for your next rocket, while you are being treated to dinner at Frere Jacques by the Solids boys, you might keep a few negatives in mind to balance all the dazzling positives. Solids tend to have lower Isp than liquids, so in general, rockets built from solids are larger and heavier. They must be handled with great care from the moment they are manufactured until they are used to ensure they are not inadvertently ignited. Because a solid rocket is already loaded with premixed fuel and oxidizer, it cannot be carried on many public roads in the US, nor shipped on aircraft or passenger carrying ships. Liquid rockets are shipped empty, weighing only 10% of their launch weight and completely safely since they carry no propellant. Solids are fueled when you buy them from the factory. Quite com-

35

monly visitors can't even enter a room storing solid propellant rockets until they have received an extensive safety briefing. Rockets using solid propellants must be built in specially equipped areas far from other people. Once they are started, solids cannot be shut off.

Because of the extraordinary care required in their manufacture and the special chemicals they use, solids tend to be expensive, though liquid rockets can also be quite expensive depending on the technological approach employed. The difference is that solid engines require highly capital intensive factories for fabricating, mixing, and bonding the solid propellant into the casing, and the propellant chemicals themselves are costly. Liquid costs tend to be in developing the engine, but recurring costs are typically, though not always, lower. Today's solids are not environmentally friendly, and their use may be curtailed in the future due to exhaust toxicity. Solids burn rapidly, leading to very high thrust levels, which is a big plus for launching, but it can also be a minus. A high thrust level subjects the payload inside the rocket to large acceleration loads. In launching from earth's surface, it is preferable to contour thrust to delay maximum acceleration until clear of the densest stratum of the atmosphere.

Got all that? If not, try the Creme Broulee. It's excellent. Make casual conversation. Here's a bit of rocket parlance you'll need. Solid rockets are called motors. Liquid rockets are called engines. Don't EVER mix those up. Like asking for La Coke instead of Le Coke, it's deadly. You might not get invited out again on another boondoggle!

Why the difference? Clearly you never studied French or you'd know the answer: Because. Because engineers, not users, dominate rocketry and they're careful people. Motor, according to engineer speak, is a prime mover that carries its own fuel/oxidizer (propellant) within it. Engine is just the device that converts propellants into motive force. Your car has an engine. Unfortunately, your grass cutter has a motor, according to common usage, which engineers would say is incorrect. Your grasscutter has an engine, too, and a gas tank. In my opinion, this definition would tell you that what makes your electric razor work is an electric engine, but most people call them motors.

Chapter 2—Propulsion

Apollo Module

Best to fall back to that rationale for why it's Das Boot, Die Fernseher, and Der Spiegel, or why it's La Mano and El Tiempo, but La Braza. Because. Solid Rocket Motor. Liquid Rocket Engine. That's the mantra. Hey, it makes them happy, and around now the check is gonna show up. Go with what works.

What Really Happens When People Build Rockets

Philosophical arguments of the relative merits of solids, liquids, Vodka, and Old Crow can bog you down, and that's not what rocketry is all about. Maybe it's most instructive to just look at what people really choose for different applications. What do we see when looking at launch vehicles and their engines? All the largest boosters, Proton, Saturn V, Ariane IV and Ariane V, Long March IV, Delta, and Titan, rely primarily on liquids, and all crewed vehicles rely on liquids. The Shuttle uses both, but the fact that a solid rocket failure caused the loss of the Challenger raised the question of the suitability of solids for carrying people into space. Certainly, for launching astronauts solids are not usable without liquids because of the high acceleration.

By contrast, all the smallest rockets use solids—Pegasus, Taurus, Lockheed's LLV series and the venerable Scout. Except for possibly the LLV series, all of these vehicles were built expressly for or derived from military missions, so their use of solids could have something to do with that. PacAstro's PA series of vehicles plans to use liquids exclusively, arguing that they achieve the lowest recurring costs combined with safety and environmental compatibility.

Most commonly nowadays, we see a mixture of liquid technology used to get high Isp and controllability along with low cost, coupled with solid propellant engines "strapped on" to provide extra thrust for initial liftoff. Europe and the US have advanced solid technologies, while the Chinese and Russians have focused on liquids, and their launch vehicles reflect these technical factors. Sometimes the first stage of the rocket is a liquid "core" with solid

Saturn V Rocket

It takes a lot of rocket to take a tiny payload to the moon.

37

strap-ons. The second stage might be solid, and the upper stages might be liquid, reflecting the need for controllability to achieve precise insertion of a payload into a particular orbit.

As if the decision between liquids and solids isn't confusing enough, a third option has been in the development of hybrid rockets. Amroc was founded to pursue this promising technology. A hybrid uses a solid, rubber-like fuel, but burns it with a liquid oxidizer, typically liquid oxygen. The hope for the hybrid is to combine the controllability and clean-burning advantages of liquids with the simplicity of solids. In fact, hybrids have proven difficult to perfect, and may bring the disadvantages of both systems into a single package. Unlike solids, they are not constantly launch-ready, and they lack the high Isp of liquids. The fluid handling equipment they require typically adds costs in ground support to liquids, while, like solids, the entire rocket must be pressurized, incurring large costs for the motor casing and for bonding the solid fuel to that casing. Liquid rocket engines are relatively small, though their fuel tanks might not be. The small engine can be tilted slightly, gimbaled, to give thrust vector control. Solids, to be primary engines, must use flexible nozzles that are more expensive and less reliable. The hybrid would require a flexible nozzle since the long, narrow casing cannot be gimbaled.

Upper Stage and High Performance Rockets

We're in orbit. But now we want to get to an even higher orbit, like the geosynchronous orbits used by some communications and weather satellites up 40,000 km (24,000 miles) from earth. Or you want to get started on a trajectory to Mars or beyond. Whatever engine you choose should have excellent Isp performance because you had to lug all of its propellant up from the ground on your primary launch vehicle, a process that cost you at least $5000 for each pound lugged.

Interestingly, you'll still find both solids and liquids up here. Solids have lower Isp, but they don't have the overhead of tankage, pipes valves, pumps, and fill/drain fittings, so for many smaller boost engines, the highly elegant solid design is best. The most famous of these upper stages is the Star series by Thiokol. These are spherical titanium shells filled with solid propellant with a partially imbedded nozzle built in. They achieve mass fractions as high as 98%, meaning 98% of their mass is propellant and the other 2% is the titanium shell, the nozzle, and some attach points—an amazing number.

Larger upper stages tend to be liquids, because as the mass of propellant grows, the relative importance of the hardware shrinks. Liquid hydrogen/liquid oxygen is the favored propellant combination, with Isp numbers way over 400 seconds. Crank that into the rocket equation, tell your accountant that propellants cost $5k per lb to lug to orbit, and this stuff looks great. Both have to be kept cold, an application of the term "cryogenic" that has nothing to do with storage of cadaver wannabes, but you have to use them pretty soon after launch. If you are liquid, but not cryogenic, you are "storable." Lots of fuels are storable, kerosene, for example.

Few oxidizers are storable because, like oxygen, they tend to be gases. Nitrogen tetroxide has lots of oxygen in its molecules, and those oxygen atoms come off pretty easily. N_2O_4, also known as nitric acid, when mixed with water, which it does readily even with any odd water molecules wafting by in the air, or the moist, cozy lining of your nose, eyes, mouth, and throat, is nasty stuff. A few other molecules are added to it to keep it from devouring the tanks it's stored in. The smoke it creates in air while producing nitric acid is a kind of witch's brew red known in the biz as IRFNA, Inhibited Red Fuming Nitric Acid. You learn to respect chemicals by seeing an open bottle of IRFNA react with plain old air.

Tiny Rockets

Once you're coasting along at GEO, or en route to your rendezvous with a planet or two, your propulsion needs are over with, right? BUZZZZZZ. Guess again. Geosynchronous orbits are not stabile. You need ΔV to stay synchronized, to keep your orbit directly over a particular spot on earth's equator. You tend to drift not only east or west of the spot, but also to pick up a bit of north/south motion. The primary lifetime limitation on many GEO satellites is that they run out of propellant and leave their workplaces. A marvelous example of our creations imitating their creators, these orbital robots spend the rest of their lives ambulating pretty much uselessly somewhere in the vicinity of their former usefulness. Like many primitive peoples, we push these feeble elders into the deserted heath. In the case of satellites, the last of their propellant is expended to push them to an orbit higher than geosynchronous where they don't interfere with the productive labors of their younger counterparts, and they spend eternity orbiting the earth.

Just how much ΔV or propellant is typically carried on board a geosynchronous comsat? One answer is: as much as possible. For many missions the propellant tank(s) are filled at the launch site to bring the satellite mass up to the very maximum the launch vehicle can carry. This can constitute half the total satellite mass. Remember the Ford Mustang we took drag racing a while back? It was also 50% propellant, and hit about 1000 meters per second ΔV with an Isp of only 150 s. Modern comsats achieve Isp over 300 seconds, thus they might carry 2000 m/s of ΔV capability to provide orbit correction and to control the spacecraft attitude.

Interplanetary spacecraft coast literally hundreds of millions of miles, aiming at targets at most a few thousands of miles across. As media coverage of these missions is wont to report, this is tantamount to kicking a soccer ball in New York through the goalie net of a soccer stadium in LA. That's not quite how it works. We kick the old ball in the general direction of south-of-west-ish, and then as we close in on California, we keep batting the ball around to a closer and closer approximation of the trajectory to the net and the final kicks are applied in the final days of a trip that lasted hundreds or sometimes thousands of days. Those kicks are applied by tiny rockets on board the interplanetary spacecraft.

In both the geosynchronous and interplanetary cases, the costs of lofting all that rocket propellant are huge. But the thrust needed is minuscule. We have plenty of time to make trajectory changes by applying gentle force over long periods, a stark contrast with launch vehicle propulsion, which doesn't need to be so efficient. For launching, you launch from the ground, and bringing propellant to

Chapter 2—Propulsion

a launch site is cheap. You need beaucoup de thrust because gravity is trying to reunite you with the ground and hence you want to get up and into orbit ASAP.

The propellants are different. The rockets are different. Even the people who build the rockets are different. Huge launch vehicle engines and motors highlight the engineering of giant structures and handling of large quantities of chemicals. I visited the plant outside of Rome where Ariane V's strap on solid rocket motors are cast. The cleaning vat used just to get the casing ready for the adhesive is 8 stories deep and 20 feet across, the world's largest washer/dryer.

But the engines that correct the trajectories of satellites are tiny. Some weigh less than 1/2 pound (220 grams). There are propellant valves whose total mass, including the seal, the actuator, and the electrical connector is less than 10% of that weight. What lengths we go to for higher Isp! Early tiny rockets used monopropellants, but now we have tiny injectors that precisely mix fuel and oxidizer into a combustion chamber the size of a thimble. Electric augmentation has been used to make the propellant warmer than just the temperature achieved by the energy of the chemical reaction. The latest trend is pure electric propulsion, which is heating propellant by passing it through electrical arcs in arcjet thrusters, some of which have double the Isp of chemical rockets. Coming soon are plasma and ion thrusters with double again the Isp of arcjets. Ion thrusters strip electrons off typically massive neutral molecules such as xenon. The ions thus formed are accelerated across an electric field to achieve exit velocities as high as 25,000 m/s.

The Final Frontier: Infinite Isp

Space is not empty, but the space near our sun is particularly filled with stuff. For one thing, it's loaded with photons—corpuscles of light—shining on us from old Sol. Our friendly star is also putting out a pretty steady flux of high energy particles known as the solar wind. The solar wind is like earth's trade winds, very dependable, very constant in both direction and strength, except for occasional

Micro Space Craft

gusts during solar storms, which would not have any significant effect on sailing conditions. Ultimately, its advantages could induce us to go solar and harness an infinite source of ΔV.

Solar sailing has been proposed repeatedly for decades. Sailing on the solar wind requires no propellant, just as sailing earth's oceans requires only the equipment and the skill to master the wind and the sea. Unlike terrestrial winds, the solar wind, having traveled almost 100 million miles to get here, is pretty weak. To get even a millionth of a pound of thrust requires unfolding tens of square meters (hundreds of square feet) of sail, and then manipulating the sail to reflect solar wind particles and push the spacecraft by means of the recoil of their momentum transfer to the sail. Unlike navigation of the high seas, where sailing predated power, solar sails must not only solve their own technical problems, but also the philosophical biases of contemporary propulsion culture, which is to carry chemicals with you and not depend on somewhat ephemeral natural sources.

Propulsion for Small, Low Cost Missions

You've been having such a great time reading all about propulsion, thermochemistry, and rockets that you almost forgot that dead-end job of yours, clinging to a few threadbare contracts for small, low cost satellites. Here are a few things to tell the boss in order to free up some overhead money for that trip to the AIAA Joint Propulsion conference in Vail or someplace.

Arguing against your major league boondoggle, bossman says, "Small satellites [Particularly the underbudget jobs you always get. Ed.] rarely have propulsion systems." He's got you there. It's fast thinking or face the ice and slush on the local ski bump. You beg for attention. Ahem, sir? He's on the phone. Subject your favorite Arbeitmeister (or Arbeitmesterin) to the following bullets, which you have committed to memory:

- One way to stay cheap is to take a piggyback ride. Those rides don't always take you where you want to go, and

Chapter 2—Propulsion

propulsion on board is cheaper than a dedicated ride into orbit.

- People are already building small geosynchronous satellites that orbit 40,000 km above the earth. At that distance over the equator, there is not enough magnetic field, nor enough variation in the field direction, to control the satellite attitude with magnetic torque coils. Small attitude control torques must be generated with small attitude control thrusters.

- Hey, people are even talking about interplanetary small spacecraft—the same attitude control problem *plus* you need course corrections en route to get to your destination.

If this doesn't get you to the deep powder, you'll have to pull out the big guns:

- All those LEO (low earth orbit) clusters of communications satellites, like Starsys, Orbcomm, Teledesic, Motorola, Ellipsat, all need propulsion to keep the satellites evenly spread out in their constellations.

See you in the hot tub for some Gluhweine!

Yes, in its innocent youth, small satellites didn't have propulsion on-board. In 1983, I was the closest approximation to a propulsion engineer in all of AMSAT (Amateur Radio Satellite Organization), compared with maybe 100 active electrical engineers worrying about radios and on-board computers. But things are changing. Propulsion is becoming a key component of many small, low cost satellites. For many of these, the combination of short mission life and low cost has resulted in use of compressed gas, usually nitrogen, which is stored in a commercially available, DoT-certified, gas bottle. But hydrazine and bipropellant commercial systems have already been flown on small satellites, and the prospects are for further increases in the number of propulsion systems on small satellites.

Chapter 3
Orbit Mechanics
—or—
What Keeps These Things Up, Anyway?

I APOLOGIZE, in this and the next two chapters, for possibly interfering with the sensibilities of those who take the subject of orbit mechanics really seriously and attack it with computers and equations (hopefully not in that order). Also, thanks to Richard Warner, who got me started on this virtually interminable subject (stay out of dark parking lots, Bub), and also made all of the original drawings that I hope clarified the hopelessly murky text. Richard also did his best to insert technical accuracy, but failures are my own. Here's our guarantee: If you find yourself lost in space due to an errant rocket burn, write the publisher for a prompt refund.

Editor's note and disclaimer of liability: This article explains the basics of what a satellite orbit is and in general terms what different kinds of orbits are available for small satellites. If any of the following describe you:

1. Busy planning a mission to Pluto at JPL using three hypercubes and two Cray-XMPs;

2. Wrote your Ph.D. thesis at MIT on nth body perturbations to satellites orbiting planets of time varying mass and mass distribution;

3. Edited a textbook on orbit mechanics or satellite guidance and control;

Micro Space Craft

4. Work as a tenured professor of Aerospace Engineering at a major university;

Then don't read any farther. Or go ahead and read for the marvelous wit and witticism, but don't write to me and complain that your brain got rusted. This article is for the rest of us who toil in the somewhat less sublime spheres or art, commerce, and left-handed thread design.

T-Off Time

Being younger than OSC's president, while he was launching monkeys I was playing golf barefoot at Lost Creek Country Club. By the time I got to monkey launching age, I had given up on rockets in favor of girls and Fords. But like a lot of kids, I too was fascinated with putting things in space. Probing questions haunted my youth, like, "If I hit the ball hard enough, will it go into orbit?" (This serious, career-minded thinking pattern, developed at an early age, explains why I exist as a ward of the near-bankrupt aerospace state while my peers are all brain surgeons and Supreme Court justices.) My older brother opined that I would destroy the ball first. Unfortunately, at age 8 you don't know how to win an argument by defining terms and splitting hairs. And that, also unfortunately, is how older brothers get away with murder.

With a real killer drive, the ball "falls" around the world.

Leave that topic for my analyst, Ernst. It turns out that my 80-yard drive into the rough *was* actually in orbit, a very low orbit that collided with the earth after only a few seconds. A really long ball sails many more seconds, maybe passing over the next hill and out of sight before finally bopping someone teeing off for the next hole ("Fore!!!!!!"). This is the trajectory marked "A" in the figure.

Chapter 3—Orbit Mechanics

Now a Truly Awesome drive (we're talking Robocop, minimum, here) could go so far out that the curve of the earth becomes important. The ball still falls to earth but a lot farther away, like Trajectory B in the figure. While you and my brother might insist this isn't IN ORBIT, depending on just what you mean by the phrase, it is AN ORBIT. When Alan Shepard was the first person to ride a trajectory like this, high above earth's atmosphere on May 5, 1961, he made quite a splash (are there no lengths I won't go to for a joke?)

If somehow you get that golf ball going so fast that the earth falls away from the ball just as fast as the ball falls toward the earth (trajectory C in the picture), the thing is not going to come back to the ground. The golf ball is in stable orbit, that is, its trajectory never intercepts the surface of the earth, unlike any of my golf balls except the one implanted in the guy teeing off at the fourth hole. Now it so happens that Yuri Gagarian rode into space on this type of trajectory on April 12, 1961, three weeks before Shepard. Sputnik did this in 1957, but the moon got into this type of trajectory around the earth quite a few years earlier than that.

Orbit Definition Good Enough for the Rest of Us

So what is this "orbit" concept? If you need to win arguments, go ahead and insist that the vase your mother-in-law gave your wife as a wedding present that had been in her family for 649 years was actually in orbit before the orbit's path collided with the tile kitchen floor after your big elbow bumped it off the counter. But at a party nobody likes a bore, and except for these special circumstances, Miss Manners would advise that you consider an "orbit" to be a more or less stable motion wherein a body continuously falls toward something big, like the earth or the sun, in such a way that its trajectory doesn't impact the big object. Examples of this highly unusual way of moving include the earth's path around the sun, the moon's path around the earth, the motions of most man-made satellites around the earth, and Marion Barry's performance as mayor of Washington, DC.

Having forgotten to also bar astrophysicists from reading this article, I will undoubtedly also be reminded that sometimes two objects go into orbit around each other, and that both orbiting bodies are of similar mass. Binary stars, they'll say. Fine. But to stay anywhere near the point of this chapter, we'll let binary stars escape my finely honed keyboard and stick with relatively itty bitty things (don't you love this technical jargon) going around big, round things (ignoring that the Kanji for big round things spells log). After all, small satellites are little bitty things relative to the big round earth around which they orbit.

How Mother Nature is Cruel to Eight-Year-Olds

There are a few problems with convincing your big brother that you can orbit a golf ball by means of a borrowed two-wood. Thanks to the pesky facts that the earth is pretty big, pretty massive, and surrounded by a viscous atmosphere (air), getting things to orbit is rough, particularly by swinging your mother's clubs choked 6" down the shaft. The slowest speed you must go to get far enough not to hit the earth as you fall is about 16,000 miles an hour (25,000 km per hr). At that speed you really don't want to be plowing through the air. The lawsuits from the sonic boom damage alone make it prohibitive, besides the fact that no known materials can withstand the aerodynamic heat created in the atmosphere at that speed. So to be in orbit and enjoy the comfort and convenience of not burning up, an altitude of at least 100 miles (160 km) is desirable.

Let's face it, eight-year-old barefoot kids wielding borrowed two-woods on Lost Creek's back 9 are not going to get a golf ball up to 16,000 mph and 100 miles. In that sense, my brother had a good point. Careers have been made on perfecting gadgets that accelerate and lift objects. People like Werner Von Braun and Robert Goddard spent their lives on the problem. Their toys are somewhat more complicated and larger than mine were. As shown below, they look like rockets, which is the generally accepted technology when you want to get to 100 miles altitude and 16,000 mph.

Chapter 3—Orbit Mechanics

V

h

For those with a bent for numbers, about 10% of the work a rocket does lifts it to 100 miles altitude (h) and 90% of the work is accelerating up to 16,000 mph (V). So you can figure that getting on an airplane going 500 mph at 8 miles altitude buys you about a 0.9% edge in getting into orbit compared to launching from the earth's surface. Thus, the real gain of a Pegasus air launch configuration is not that it's a good way to get into orbit, rather that it's nice to be able to leave the ground from an airport instead of a launch pad.

At 100 miles, moving at orbital velocity, one orbit around the earth takes about 95 minutes. One idea for very fast transportation is to make an airplane that can achieve a near orbital trajectory. While not capable of reaching a stabile orbit altitude and velocity, it would, like trajectory B, "fall" almost half way around the world in under 50 minutes. This would mean London to Los Angeles in 35 minutes (with airline flights that fast, my writing would never get finished — but my laptop's battery-life would finally be near-adequate). Another major minus of suborbital travel is how painful it would be to spend an hour waiting for luggage at Dulles after coming in from Tokyo in 40 minutes.

Rockets that now provide this "suborbital" service are called sounding rockets, probably because they have been used over the years to take upper atmosphere soundings. Usually carrying an instrument payload to altitudes around 100 miles, they do not achieve full orbital velocity and hence their fall intersects the earth's surface. They can provide exposure to space vacuum and zero gravity conditions for about 10 minutes before reentry.

More suitable toy than a 2-wood for pursuing a stable orbit by getting 100 miles up (h) and accelerating to 16,000 mph (V).

Micro Space Craft

Orbit Altitude

I rode on the good ship Susquehanna recently, or at least its replica, which plies the harbor of Izu peninsula, Japan. ¥300 is all it takes to tour the harbor where Commodore Perry landed. And a hundred miles is all it takes to be in orbit. But who wants to tour the famous scenic harbor from inside the hull of the replica Susquehanna watching the paddle wheels go around? You want to be up top, breathing the air, taking pictures, and feigning throwing your friends overboard. To do that costs ¥250 extra, payable at the candy counter once the boat gets moving.

A hundred miles means you're in the crew's quarters of orbit. For one thing, there's still a little air up there. Not a lot, but at 16,000 mph you hit the few molecules there pretty hard and pretty often. The drag slows you down and pretty soon your velocity isn't high enough to keep falling past where the earth's edge is and, like that old vase, you are in for a landing with the earth, possibly on the kitchen tile floor. Luckily the atmosphere's density drops off exponentially. Thus at 100 miles altitude you're going to stay in orbit for a week or two, but at 175 miles, where the Space Shuttle usually hangs out, you'll last a few months. The Shuttle stays on orbit only about a week, not because of orbit mechanics, but mainly because the number of Disney videotapes that can be stored on board is limited and the crew gets cranky. At 300 miles, orbital lifetime (the jargon to use at cocktail parties once you've identified the particular creature you want to mesmerize with your command of aerospace engineering) increases to about 10 years. Just a little higher, like 500 miles, and your great-great-great-grandchildren are the ones to worry about where your primitive satellite smashes on its return to earth.

Getting up high has a big advantage in lifetime. Unlike riding the Japanese Susquehanna, the relative cost to go higher, once you've paid the basic ¥300 to get to 100 miles and 16,000 mph, is a very small additional amount of energy. So why not always go higher? Well, if you want to get every last ounce (38.38 grams) into orbit, you can lift more to a lower altitude. Also, the radiation dose people and electronic things receive goes up a little with altitude. Plus, if

Chapter 3—Orbit Mechanics

you want to make color glossy pictures of the shifting sands of the Sahara, your friends' ABM launchers and license plates in Moscow (Why do they always boast that some $1B satellite can do this? My mother-in-law did it on vacation in Russia last year with a cardboard camera she bought with the film built in for $17 at K-Mart). Well, anyway you can get better pictures of the earth from lower down. Some people like to experiment with the chemicals at the top of the atmosphere. And some people are just bad calculators or they ran out of money building a rocket that was supposed to go somewhere useful and just ended up at 100 miles.

Another advantage of being higher up in orbit is that you can see points on the earth that are farther away. Most spacecraft radios are "line-of-sight" like TV and FM radio signals, and by getting the satellite up higher the satellite can reach a larger region, shown in the figure.

A bit above 500 miles you enter a region of increased radiation intensity called the Van Allen belts, which extend to about 5,000 miles. You don't find too many satellites in this region. Matter of fact, the next popular orbit altitude is about 24,000 miles (40,000 km), a long distance, being about 2 earth diameters away (or 450 times farther than the distance from Toledo to Lima, Ohio). The ticket to put your satellite in 24,000 mile orbit is 3 to 5 times more expensive than a low earth orbit launch. Signals are 6,400 times weaker from 24,000 miles than from low earth orbit satellites at 300 miles. The exquisite radio antennas used to reach earth from that distance make that coat you bought your wife to get her over the loss of the crummy old vase look like pocket change, not to mention the tracking system you need to point those shiny antennas at the ground station.

The higher you are, the more you can see.

Micro Space Craft

So What's Everybody Doing Up Here?

Having already covered about 10,000 years of what humans have learned about orbital mechanics, it seemed like a good time for a pepperoni pizza (lots of small round shapes with big round shapes, according to Ernst). Pick up the phone, order one or two, and prepare yourself for geosynchronous orbits, Molniya and other eccentric orbits, sun synchronous orbits, what orbits look like from the earth, and some special tricks like earth escape and minimum sunlight orbit. Please, if the pizza does not supply enough heartburn, then while finishing it, think about the following homework problem:

The moon is smaller and less massive than the earth and has no atmosphere. Could you get a golf ball into orbit there? Assume Cher gets rid of her perm and Microsoft elects to build a sprawling complex within the lunar surface instead of the Lunar Towers IV Center now in preliminary design (but unlikely to gain lunar zoning approval), so very tall obstacles should not be an issue.

Chapter 4

Orbit Mechanics II: The Movie

THEY SAY the line between genius and insanity is a fine one. Having puzzled over the quiz provided at no extra cost as part of this nail-biting account of orbit mechanics, you're probably wondering what the call is. Well . . what was the question? How hard is it to put a golf ball in orbit from the moon's surface?

The good news is that with no atmosphere and assuming a spherical moon, meaning no mountains or office towers poking up above the ground, you don't need to provide energy to gain altitude. How fast does the ball need to go to reach a stabile orbit scraping the surface of the moon? The moon's mass is about 12% of the earth's, and its radius is 27% as large. This means lunar orbit velocity is about 21% of earth's, which would make it 1,550 meters per second (about 3,465 mph). For comparison, a golf ball hit on earth with that huge velocity off the tee would travel, ignoring aerodynamic drag, about 79 miles. With drag, maybe 10 miles. So if you're the kind of guy who tees off at the Los Angeles Country Club and chips onto the green in Palos Verdes, you have a career ahead of you in golf ball launching on the moon.

The answer to the quiz is, sadly, no, not with a golf club. With a small rocket maybe. If you want to put things in orbit with a golf club, find a small asteroid where gravity is real weak.

Now that we're all relaxed and resigned to carry on with our mundane lives instead of putting golf balls into zero altitude orbit around the moon using golf clubs, we can attack some pretty substantial issues in satellite orbits like:

— Geosynchronous Orbits (GEO)
— Sun Synchronous Orbits
— Orbit Planes
— Ground Coverage.

Geosynchronous Orbits (GEO)

Probably the orbit most important to The Rest of Us (the people this article was written for) is geosynchronous orbit because satellite transmission of TV, radio, telephone, and telemetry (data) uses satellites in this orbit. This wasn't always true. Back when televisions used to come in something called black and white and phones used to have something called dials on them, and people used to cook food by heating the air around the food in big insulated boxes and listen to music made by the vibrations of small needles travelling in plastic grooves on rotating discs, back in the time before any of today's college students were born and I was playing the back nine of Lost Creek in peace, a few satellites were used for television relay from low earth orbit, similar to the orbits ridden by Yuri Gagarian and John Glenn. These orbits pass over any point on earth very quickly since they travel around the earth in an orbital period lasting about 95 minutes. This short time period is inconvenient because to transmit a TV broadcast via satellite between two places on earth, both places must be able to look up in the sky and "see" the satellite with their radio antennas. In a low earth orbit, this visibility period is short, typically five or six minutes. If this were all we could do with satellites, there would be no Bart Simpson. Let the gravity of that statement settle in for a while, eh? Well, there are two solutions. Geosynchronous orbit is one of them.

Wouldn't it be nice to put a satellite in orbit that doesn't circle the earth, but rather just "hangs" over a single spot on the earth's surface? Such a satellite would enable 24 hour a day TV transmission—Bart included. Gnarly, you might say.

Chapter 4—Orbit Mechanics II

Not to throw cold water on a good idea, but in general, a satellite that just hangs there is impossible. I distinctly remember mentioning that orbiting bodies can be thought of as continuously falling toward earth. The reason they don't simply fall to earth, making a large crater but not really doing what they're supposed to do, is that they move forward fast enough to fall over the arc of the earth's horizon. Two facts of adult life are now going to be exposed to the reader. If you are under 13 years old or if you are one to cling to faint hope, stop reading now.

    1. There is no one named Santa Claus living at the North Pole. Nobody lives at the North Pole.

    2. Satellites don't "hang" up there. (reverent silence)

Let's say we're in Low Earth Orbit (LEO) over the equator flying west to east (As the World Turns) at 350 miles (600 km) altitude. Every orbit takes 98 minutes during which time the earth turns a little, and it turns out that it takes 105 minutes to pass over the same point on the ground. What if we raise the orbit altitude a little? The orbit is now going around a bigger circle, and the earth's gravity is weaker because the distance to earth has been increased. The earth takes a whole year to get around the sun because we're so far from the sun that the path around it is a long way, about 560 million miles.

The combination of weakening the pull of earth's gravity and increasing orbit circumference with altitude means that orbit period increases with orbit altitude. At 24,000 miles (40,000 km) the orbit period is 24 hours, which is the same as the earth's rotation period. The figure on the next page shows that a satellite in orbit over the earth's equator at 24,000 miles still goes around the earth, but at the same rate the earth turns, so the satellite appears to move in synchronized motion with the earth—hence the term geosynchronous (or, in the case of this particular illustration, Eiffel-synchronous).

There are a lot of nice things about geosynchronous satellites. They appear stationary relative to us earthlings and our backyard satellite dishes. Once you get the dish planted on the ground, it doesn't have

Micro Space Craft

to move to track the satellite's motion. Satellites in geosynchronous orbit are high enough to see half the entire earth at once, like the Apollo mission pictures of earth showing it as a big circular blue, green, and white disk. Seeing this much means that a satellite can relay TV, telephone, and other communications over long distances in a single hop.

The few disadvantages of geosynchronous satellites are due to the simple fact that they are far away. Traveling at the speed of light (or radio waves), it takes about 1/4 of a second for your electronically transmitted voice ("Mom, Sheila and I are getting married.") to reach from your condo in Culver City, LA to your parents' condo in Sarasota, FL. Add another 1/4 second or so for their voices to answer you (Mom: "My Baby!"; Dad: "Sheila who?"). This delay is doubled if the conversation is carried across two geosynchronous links, which can happen when calling, say, from LA to Tel Aviv. As difficult as this can make conversation for you, it's really hard on your computer's modem or your fax machine, because these devices rely on constantly hearing replies from the other end of the line. They are designed to wait for these replies in cases of distorted transmission.

From a long distance, radio signals become pretty weak, mainly due to the beam spreading, just the way a light appears dimmer seen from a distance. Elaborate, expensive antennas are used both on the satellite and on the ground to reduce this spreading. The high power transmitters needed on the satellite despite these antennas are also expensive. All the fuss about direct broadcast satellites is based on using more powerful satellite transmitters and narrower beam satellite antennas so that you don't need a huge dish antenna in your yard to receive satellite broadcasts.

A geosynchronous orbit maintains the same relative orientation relative to the earth, and the Tour Eiffel, throughout the day.

Chapter 4—Orbit Mechanics II

I mentioned that there are two solutions to the problem of using satellites in low orbit for TV transmission. One is to raise the satellite orbit to geosynchronous orbit (GEO). The other is to fly many satellites at low orbit arranged so that one is always in sight of you and one is always in sight of the stations you're communicating with. Since they don't require such elaborate transmitters and antennas, these LEO satellites could be much more simple, and that is why small satellites are being proposed for LEO networks. The disadvantage of the approach is you've got a lot of satellites moving in their own independent orbits, and a lot of "cross linking" is required. Satellites at LEO see only a small part of the earth's surface, so to communicate over long distances, the signal is unlinked to one satellite of the system, crosslinked among satellites, and finally downlinked by a satellite in view of the receiving station. A further complication is that the constellation of small satellites moves relative to observers on earth, and hence the uplink, cross link, and down link satellites are constantly changing. (See Chapter 16 for more on how LEO satellite clusters work.)

But let's leave the world of telecommunications to the telecommunicators and get back into orbits. Orbits are described in part by their inclination, meaning the angle the orbit makes with the equator. Geosynchronous satellites are normally placed in equatorial orbit because they can only be synchronous with a point on the equator. This is because every orbit makes a complete circle around the earth. Further, this circle has to lie in a flat plane that intersects the center of the earth. This is because all objects fall toward the center of the earth (which is why people think "down" is such a simple concept.)

Thus, for that "geosynchronous" orbit altitude over the North Pole (yes, where Santa Claus doesn't live), the satellite would have to proceed south, pass the South Pole and reappear over the North Pole 24 hours (1 orbit period) later. Not a tremendously handy orbit, at first glance. But in low earth orbit, being at 90 degree or "polar" inclination has a lot of interesting advantages. A satellite in orbit inclined 90 degrees to the equator at 350 miles (650 km) "sees" every point on earth twice a day. This orbit is show in the figure below.

Micro Space Craft

Every 95 minutes the satellite makes a circle over the earth following roughly a line of longitude on the globe. But by the time it returns to the North Pole, the earth has rotated about 1/18 of a revolution. Thus, on the next orbit, it follows a new longitude line about 1000 miles west of the previous orbit path. Imagine a satellite in polar orbit that passes over Cleveland at noon. An hour and a half later it passes over Denver. An hour and a half later, three hours after passing Cleveland, it passes over San Francisco, where it is local noon. By carefully selecting orbit altitude, it is possible to put a satellite in an orbit that makes exactly 1 or 2 or 3 or 4 or ... 18 orbits in 24 hours. These orbits pass over the same spot on the earth at the same time every day. Except for one small problem. The satellite's orbit is in no way linked to the earth's rotation. It is pretty much stationary relative to a fixed reference frame, like the stars and the sun. So as the earth revolves around the sun, the synchronicity is lost (but a Police album is gained).

The orbit path remains (to first order) stationary while the earth rotates.

Fear not, those who want their own satellite to pass over their house every day at 5:00 a.m. and thereby to own the world's most expensive alarm clock. There is a special orbit called a sun synchronous orbit, which is shown below.

This orbit is inclined slightly away from pure polar orbit in such a way that the orbit plane (the imaginary disc whose edge is the satellite's path) rotates (precesses, if you keep that kind of company) once a year and the satellite stays sun synchronous indefinitely.

Since so few of these satellite alarm clocks have been sold, who uses sun synchronous orbits? Imaging satellites use them. They can compare traffic patterns or crop growth or nuclear submarine factory construction best by imaging targets at the same time of day every day, to get the same sun "color" and the same shadow patterns. A neat aspect of the sun synchronous orbit is that if you're synchronous at the right time of day, the satellite almost never goes into the shade of the earth (Can you say "umbra?" It sounds very chic.) Note

Chapter 4—Orbit Mechanics II

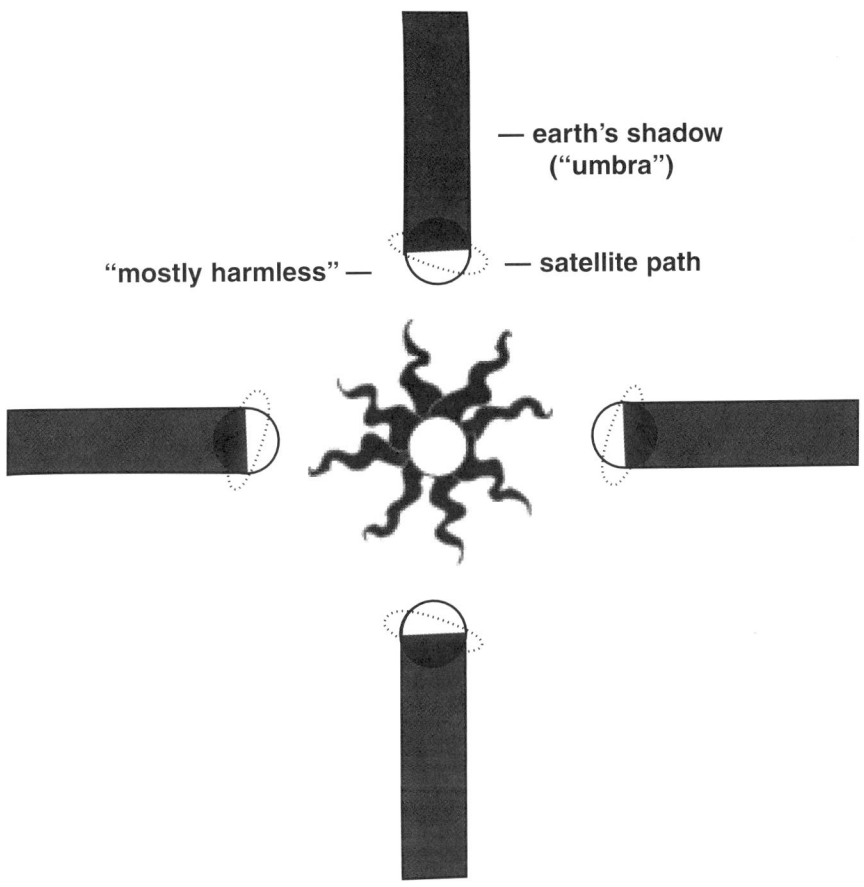

A sun synchronous orbit maintains the same orientation relative to the sun throughout the year.

in the figure that even though the satellite flies over shaded parts of the earth (areas where it is sunset or sunrise), it is high enough that the satellite itself can always see the sun. This means the thermal environment of the satellite can be very stable and it may need few or no batteries to power electronics since its solar panels are almost continuously illuminated (batteries might still be needed to handle peak loads).

Well, I don't know about you but all this satellite talk is really tiring and I'm ready to pop open a cool one and flake out in front of the

Micro Space Craft

tube with the Simpsons (try that one on your Japanese tutor). But that fat lady, whoever she is, is not singing until we get to talk about getting from one orbit to another.

Chapter 5
You Send Me: Orbit Mechanics III

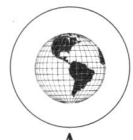
A
Initial Orbit

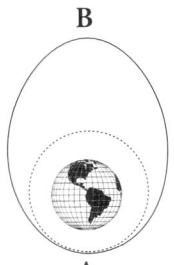
B
A
Transfer Orbit

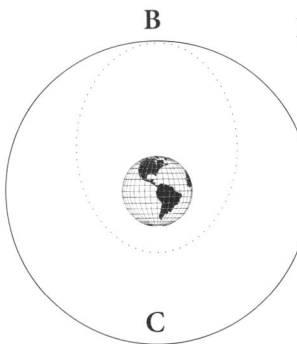
B
C
Final Orbit

A Hohmann transfer orbit is the most energy efficient way to go from one circular orbit to a different circular orbit.

MY WIFE, Nancy, is always explaining to me that particles, satellites, cars, and other so-called inanimate objects do not have feelings, ideas, and aspirations. She calls them "it" or "that." I call them "he," "she," or "that guy." While most people find our points of view irreconcilable, I believe the only difference is that she never owned a 1969 MGC. Talk about personality (and usually a pretty conniving one at that). I promise not to depart on a musical interlude into how this car knew when it was being neglected and how it arranged for suitable revenge. Rather I mention this because in at least one sense she (Nancy) is right. Simple objects don't know much about where they've been or where they're going. Like horses, they live for the present, and are probably much happier for it.

This chapter, for those of you stepping into the middle, which means everyone since I started in the middle and intend to stay there right up to the end, is about modifying satellite orbits. Three such orbits are shown schematically on the left.

The innermost one is like a low earth (low, circular) orbit (LEO), the outermost one is like a geosynchronous (high circular) orbit (GEO), and the elliptical one is an elliptical orbit (breakthrough insight, that was).

61

Micro Space Craft

The elliptical one is more often thought of as a geosynchronous transfer orbit (GTO). The two circular orbits could also represent the orbits of two LEO satellites, one higher than the other, as for instance the Space Shuttle in low orbit and the Hubble Space Telescope higher up. What follows then is the story of how to get up and down.

Imagine you're one of two satellites, both now at point A on the diagram. One of you is in LEO and the other in GTO. How do you know which of you is which? At A, you are both pointing in the same direction, but the one in GTO is going faster than the one in LEO. This extra velocity makes your fall toward the earth less steep and you swing farther outward.

So now you're at point B, meeting up with another orbiting friend. Again, how do you tell who's in GTO and who's in GEO? As before, your friend in GEO is going faster than you, albeit in the same direction, at point B. Hence you fall a little more steeply toward the earth heading for your rendezvous with point A again.

This simple revelation doesn't tell you everything you'll ever need to know about maneuvering in orbit, unless you never planned to go there anyway, which is, after all, sort of the degenerate case. But it does tell you a lot. To get from LEO to GTO, all that's needed is to align a rocket motor with your local direction and accelerate to the velocity of a GTO orbiting body at its perigee (orbit low point). This is the first step in reaching GEO from, say, the Space Shuttle that drops things off in LEO. Note that wherever you burn your engine (point A), you return to over and over. Rocket burns at perigee raise the apogee (orbit high point), but only affect the velocity at perigee.

This maneuver is sometimes called Apogee Raising, but in reality you are Perigee Accelerating. To circularize into the higher, circular orbit, a second rocket engine burn, called Perigee Raising, is effected at the apogee.

These maneuvers can also be done in reverse by slowing the velocity. At some point in a circular orbit, the orbit becomes elliptical

Chapter 5—Orbit Mechanics III

with a lower perigee. This is how people get themselves out of orbit and back to earth.

All of this maneuvering by changing the forward velocity to raise and lower apogees and perigees is generally grouped into the term "Hohmann Transfer" because Professor Hohmann first suggested it and calculated how much velocity change is required to effect a particular apogee or perigee change. There are other maneuvers, but this one is really the Big One. It has applications beyond just going from the Shuttle to GEO. For instance, say you're a satellite at GEO, and you're watching the weather over the West Coast of the U.S. and the Pacific. And your friend, a satellite whose job is to do the same thing over the U.S. East Coast and Atlantic, dies tragically of a burned out light bulb (don't laugh; that part actually happened). To make matters worse, it's late summer, boring weather time over the Pacific, but hurricane season in the East. So you want to hustle your buns on over to the East and fill in for your departed amigo. How to do it?

Remember, the periods of orbits are shorter as the orbit altitude (radius) decreases. For instance, it takes the earth a whole year to get around the sun, but it only takes the moon a month to get around the earth (it's a lot closer than the sun), and it only takes a satellite at 750 km (400 nautical miles) about 100 minutes to get around the earth. Geosynchronous satellites, by comparison, take 24 hours to get around the earth, which is why they are called geosynchronous. (For that matter, the fact that the moon takes a month to get around the earth is why the time period is called a month, or why the moon is called the moon, depending on your point of view.)

But I digress. To move from synchronous orbit over the equator below the U.S. West Coast to the equatorial position south of, say, Bermuda, you need to speed up relative to the earth; that is, get into an orbit with a period of less than one day.

No Problem! says Prof. Hohmann. Just do a perigee lowering maneuver (burn your rocket to slow the satellite's velocity). You'll drop into an elliptical orbit with the apogee unchanged, but a lower perigee, and that new orbit has a shorter period, meaning that it goes around the earth in slightly less than one day. So relative to an observer on earth, you'll slowly move West to East across the sky. When you get to the eastern location you want, you again burn at apogee, this time in the prograde (as opposed to the retrograde) direction, raise the perigee back up to equal the apogee, and you are ready to watch for hurricanes.

This use of partial mini-Hohmann maneuvers is the way geosynchronous satellites keep themselves exactly over their desired locations. It is one of the two major reasons geosynchronous satellites have to carry a significant amount of propellant, and exhausting that propellant limits the useful life of the satellite.

The other reason these satellites need propellant is that geosynchronous only means something over the equator. Remember Santa's geosynchronous polar orbit. You'd pass over the North Pole once every 24 hours, over the equator on the way south once every 24 hours, and so on. But you would not appear stationary in the sky. What do polar geosynchronous satellites and jolly elves with beards living at the North pole have in common? They would both be very useful, were they to exist, which, Virginia, they do not. Geosynchronous satellites, besides needing to trim their orbit altitude, need to trim their orbit inclination to stay truly geosynchronous.

Imagine the erstwhile equatorial geosynch orbit somehow gets inclined by one degree, oscillating above and below the equator once a day. All the ground antennas would need to move around to track it. Since geosynchronous satellite dishes are not outfitted to be moved easily, this is a nuisance.

The last gasp of this prolonged treatise into satellite motion, then, is how to change planes, like go from one degree inclination to equatorial (0 degree) inclination, or go from equatorial to polar. The problem is the same. Going back to the line of thinking that satellites are rather naive when it comes to remembering and planning, how does a polar-orbiting satellite in 24-hour orbit (high up) know it's any different, as it crosses the equator, from another satellite in

Chapter 5—Orbit Mechanics III

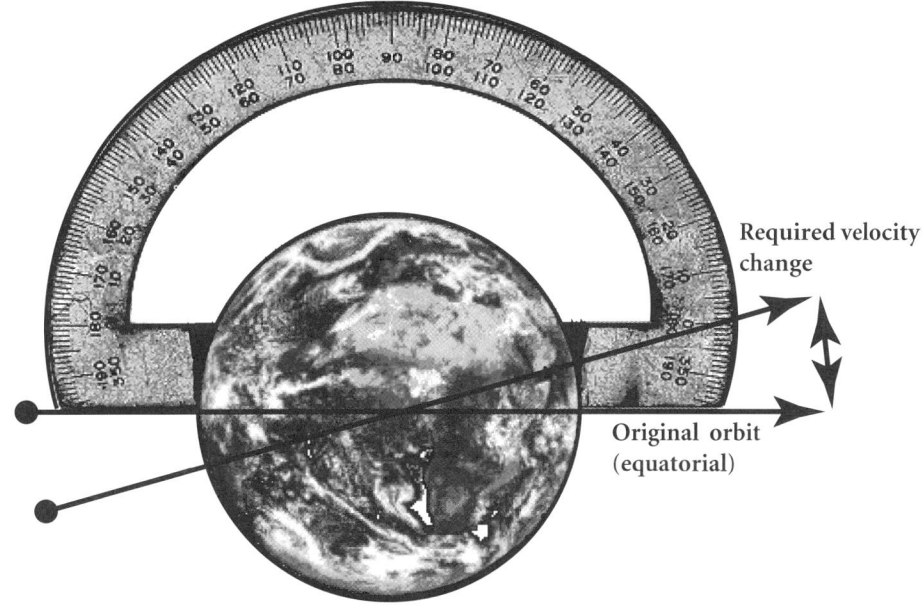

Small change in inclination requires small velocity change

geosynchronous orbit at the same place? Its velocity is different. Period.

So as the geosynchronous-altitude-but-polar-orbiting satellite crosses the equator, he (or she) needs to use its rocket to provide enough velocity increment to cancel the north-to-south motion and add an equal amount of east-to-west motion.

Radically changing planes requires a truly huge amount of velocity increment (thousands of meters per second or miles per hour), so nobody does it because they can't afford the propellant. But to go from 1 degree to 0 degree inclination requires only about 2% as much velocity change as making the radical 90 degree to 0 degree change. But even a single degree is still a significant cost to the propellant reserves of most satellites.

The maneuver is: wait until you are crossing the equator (say north to south), and propel yourself northeastward to both cancel the

Micro Space Craft

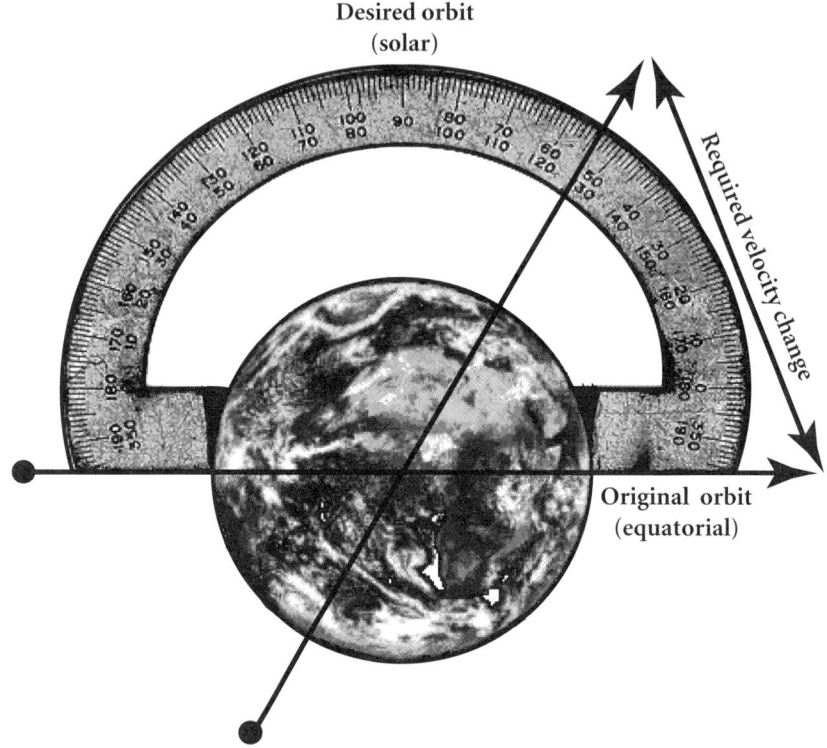

Large change in inclination requires large velocity change

southward component of your velocity and to replace it with an eastward component.

Which brings us to the conclusion of this odyssey in orbit mechanics. Ernst (my in-house Jungian phenomenologist) has warned me that a deep depression might set in once I don't have this project to occupy those lonely hours on airplanes and in airports and at meetings on totally unrelated topics where my incessant typing on the old Cambridge Sinclair Z88 is thought to be careful note taking, but is in fact my communion with orbits and readers. In fact, the earth's ecology will improve slightly as my AA battery consumption will go down. There's little compensation for you, but I'll be fine, occupying myself doing what other great men before me have done—leering at young girls, drinking to excess, and taking drugs in hotel rooms.

Chapter 6
Magnetic Attractions
—with Richard Warner

REN
KIN
JUTSU

Introduction to Alchemy, Magnetism, and Cold Fusion

GOLD IS A MALLEABLE, nonmagnetic material. Unlike other metals such as iron, chromium, and aluminum, it is pretty much useless for building things — too soft. Nonetheless, it has been highly valued for thousands of years because it is beautiful, rare, and does not tarnish. Relatively recently we have noticed that gold is among the best conductors of heat and electricity, making it the material of choice for the fine wires of many integrated circuits and for coating electrical contacts. In the West we associate alchemy—a mix of chemistry and magic—with the production of gold from base materials like lead. But the ancient Chinese also knew gold well enough to describe the magical field in three simple characters that mean the art of tempering (hardening) gold, a process about as plausible as gold's production from lead.

Today's engineers, of course, would resent labeling their work as alchemy. We prefer to say we understand an analytic basis for the way things work. This whole book implicitly rests on that questionable hypothesis—that we can explain what we observe, and we understand the principles of what's possible and what's not.

What about cold fusion? First it was impossible. Then Congress was considering putting $50M or so into Utah to pursue it, and Nobel Prize winning physicists advanced new theories on how it could be. Today fusion in a teacup continues to occupy a niche in the scientific underground, suspended somewhere between understanding (which equates to acceptance) and rejection. Its disciples carry on their research, books are written on its remarkable promise for earth's inhabitants, and others are written on its impossibility. So do we understand the possible and the impossible? Is cold fusion the bumblebee of the '80s and '90s—the flying insect we can prove can't fly?

When in doubt, change channels, I say, and with 500 of 'em out there, why not? Most of what I know about nuclear physics I learned in Isaac Asimov books, and my considered opinion on Cold Fusion is that if other people understand that stuff so well, how come they can't settle their arguments about it? I also know pretty little about changing lead to gold, or the fact that gold can't be tempered, but I'll accept the word of the experts that it's not too practical a thing to do.

Where do magnets come from? If cold fusion and making gold out of recycled bottles were all the alchemy we had to deal with, I'd dismiss it. But then there's magnetism. Frankly, few engineers really understand it. Light and radio we understand; they're photons and wave functions. But the typical explanation of magnetism goes something like this. Why is a bar magnet magnetic? Because iron atoms each have little magnetic dipoles (think of them as tiny bar magnets) in them. Normally these atomic magnets are randomly oriented and cancel each other out. In a magnet, many of the iron atoms that make up the bar have their magnetic dipoles lined up so that they don't cancel out. And voilà, a magnet.

I got this rather unsatisfying message at Wiley Junior High School in 1967, the same year I was sent home for dress code violation

Chapter 6—Magnetic Attractions

despite my conservative button down shirt, pressed navy pants, dark socks, and penny loafers (with pennies, btw), when my personal article-left-somewhere-in-the-gym-locker-room du jour was my narrow leather belt. Empty belt loops were not acceptable in public school. Period. Sending a little kid out for a three-mile walk home and back in subfreezing weather to correct the deficiency was totally PC. I like to tell that well worn tale to the odd 12 year old to permanently gel our generation gap. They like that.

Anyway, it was in '67 that I got a feeling I wasn't going to get along with magnetism (or nice clothes) at all. Frankly, both have been trouble to me ever since. Dressing for success I just can't handle, and what's the alternative? Dress for failure? The Dress for Success paradigm doesn't leave any room for dressing for comfort, variety, speed, economy, warmth, or even utility. It's success. Or failure. Tough choice. Magnetism-wise (in the '50s and part of the '60s people used the -wise suffix a lot, along with -nik as in beat-nik, kibbutz-nik), I had this little proto-conversation (the kind you don't ever actually have, you just argue with yourself and get totally frustrated with the other person over, when in fact you haven't said a single word to them) with my well-meaning science teacher, who really knew something about biology, but unfortunately not much about physics: "Excuse me, Mrs. Wolf (we said Miss and Mrs. back then and didn't think anything about it. Today you just say excuse me with a subtle hard edge, and end your phrase as if it's a question, as in "Excuse me, I don't think so(?)" which translates as "You are an idiot".) Magnets are explained because of tiny magnets? What about the tiny magnets? Where does their magnetism come from?

Oh, I got an answer. About 7 years later in college physics. I remember distinctly that the blackboard was filled with vectors and tensors, remarkably similar to the vectors and tensors filling the chapter in the textbook on Magnetism. There are all these quantum variables which, if you accept them as God's truth, and if you then invest a few years in learning the math to manipulate them, convince you that there is a fundamental principle lurking behind magnetism. The tiny magnets were just a bit tinier in college than they were at Wiley. At least the collegiate experience allowed me to lose the belt without consequence.

What Rocks Tell Us

Everyone should launch a satellite at some time in their lives that doesn't work. Known in the business as "a rock," you learn a lot from an information machine in orbit that cannot transmit any of its information—the tree falling in the forest of the info age. You learn empathy for Woody Allen, particularly that short flick where his overbearing Jewish mom dies, but ends up as a giant face in heaven staring down at him 24 hours a day. On dates, at work, at the grocery store, his Mom's huge face in the sky constantly mocks him. Dear, your glasses are smudged. Dear, you'll never get anywhere with that attitude. Dear, you were so cute as a little baby when I changed your diapers on the kitchen table. Etc. Your rock, orbiting over your head in its total, multimillion dollar uselessness, teaches humility. You may think you know how to build things, and how they work, but you are just a little bit wrong.

Geologists don't launch rocks, but they live with them. One of the most significant rocks they live with is the earth, which happens to have a huge magnetic field. Why? It is a testimony to our lack of understanding of magnetism that nobody knows why the earth has a magnetic field. Oh, it has something to do with the earth's molten iron core. Molten iron, however, has been available to us humans for at least 10,000 years, and maybe 100,000, and nobody has ever made a magnet out of it. In fact, one sure way to un-align all those tiny magnetic dipoles and demagnetize iron is to heat it, particularly to melt it. Magnetism depends on structured order of atoms, something liquids ain't got. Maybe it's the molten core being spun around and circulating in all kinds of nifty ways due to Correolus accelerations, gravity, thermal gradients, and ferrous boundary conditions from the solid phase crust? Sounds like "Return of the Really Tiny Magnets" to me.

Now, if you accept those bizarre quantum variables, which turn out to not even explain all of what physicists already know about matter and energy, but do explain some other things that physicists know about matter and energy, and you learn the math, and then you accept the molten iron, Correolus, special boundary conditions and scale effects stuff, you are going to be totally comfortable with the earth's magnetic field. I just prefer discomfort.

Chapter 6—Magnetic Attractions

Magnetic Many Uses Game

Maybe it's just as well magnetism is a bit magical. This would explain our contemporary usage of it—he has personal magnetism, for instance. We don't know what personal magnetism is, but we know it when we sense it. So let's agree that magnetism is like music. We really have no clue how it works, or why we like it, but we do, and we find all kinds of neat uses for it—taking dates to rock concerts, masking awkward silences in elevators, stimulating us to buy more in grocery stores, etc. In the case of magnetism, particularly aboard satellites, there are really just two uses: figuring out something about your position and/or attitude being the more familiar one, and doing something about your attitude being the other one.

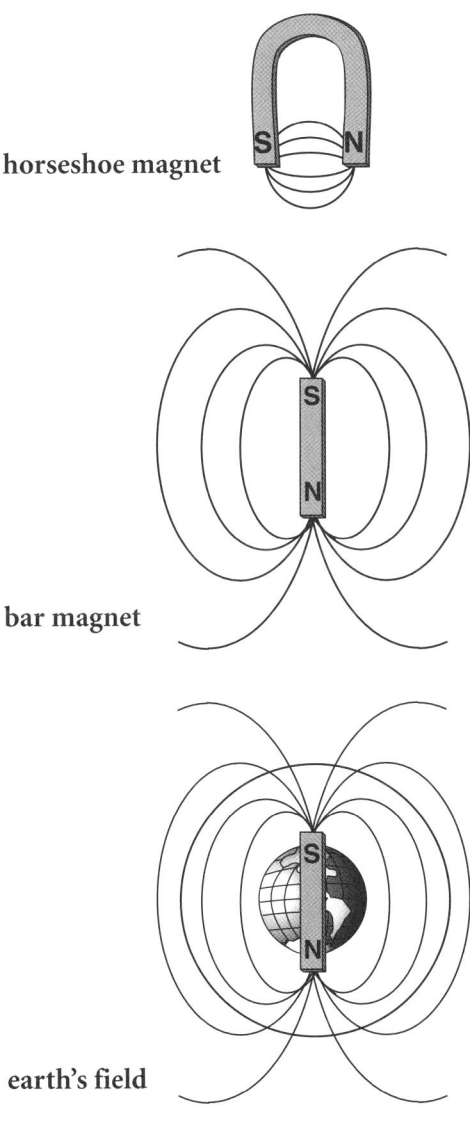

horseshoe magnet

bar magnet

earth's field

Magnetic fields I have known.

Both functions are quite practical to do at the low orbit altitudes most small satellites use. The picture shows that you are just as conveniently serviced by earth's magnetic field at LEO (low earth orbit) as you are on earth's surface. And since LEO satellites move around the earth quickly (an orbit about every 100 minutes), they see a lot of variation in the field direction. We'll soon see that this variability of the perceived field is particularly convenient for attitude control.

Satellites in very high orbits, like GEO (geosynchronous, 24,000 miles up), are too far away to see much of a field at all. And because they orbit very slowly, and in synchronicity with earth's own rotation, they see almost no field variations. Thus, navigating and steering via the earth's magnetic field is impractical for them. The situation is even worse magnetism-wise for interplanetary spacecraft, which see virtually no external fields except if they happen to be passing by Jupiter, or if they have very sensitive instruments to detect the minute magnetic influence of the sun.

This simplest model of the earth's magnetic properties (shown above) is a large bar magnet with its south pole in Canada and its north pole near Antarctica. It's a confusing convention, because the north magnetic pole of the Earth roughly corresponds to the south geographic pole. However, it makes all other magnets consistent; the north pole of a magnet wants to point geographically north.

Assume a Can Opener

It is handy to think in idealizations. Easterners like to think, for instance, that California is a Shangri La of beautiful weather. This image gives them something to aspire to throughout their lives, that they might eventually retire to some sleepy seaside community, soak up the sun, and relax. They skip over the crowding, obscene housing prices, earthquakes, floods, fog, rocky beaches, cold water, air pollution, skin cancer, and Mediterranean fruit flies. Or we want to think of our government as a benevolent and intelligent organization that ensures our personal safety, educates our kids, and protects us from poverty and disease in case of hardship. See how nice those idealizations are?

Here's another. Heat transfer is extremely dependent upon geometric factors. The only heat transfer problems people can actually solve when they learn heat transfer, involve the temperature profiles of objects with geometric simplicity like infinite flat plates and perfect spheres, though not too many of either are dotting my landscape. The perfect sphere is used to idealize the shape of a Thanksgiving turkey, and every budding engineer goes on to calculate the cooking time of a 14 lb. bird. Whether you assume a mythical man-month, a spherical turkey, a rational, benevolent government, or perfect markets in equilibrium with access to perfect information, idealizations are abstractions that help us, but can also delude us.

I hope the material above will attenuate any great stress placed upon the reader who now learns the shocking truth that the earth is far

Chapter 6—Magnetic Attractions

from a perfect magnetic dipole (i.e., idealized bar magnet). The field has many fluctuations around it that render it quite unsymmetric. In fact, it varies from year to year, too. Mapping the earth's magnetic field is an ongoing activity that creates some job security for a few small satellite builders who need to put a satellite up there, know where it is as a function of time and which way it's pointing, and send to the ground the readings from an on-board magnetic field sensor, a magnetometer.

Satellite Compass: the Magnetometer

The most common magnetometer is called a flux gate magnetometer. Owning one of these used to brand you as a real techie since they were complex and pretty expensive. Nowadays, low cost flux gate magnetometers can be purchased as an option on your Minivan, or inside of a digital watch, right alongside an integrating altimeter. If Columbus had one of those watches, Manhattan might have ended up in the Indian Ocean and Arthur C. Clark would be e-mailing his best sellers to a publisher in Calcutta from somewhere in the wilds of Long Island.

The advantage of a flux gate magnetometer over a scout's compass is that it tells you the field in 3 dimensions quantitatively. Or put another way, it describes the local magnetic field line to you in terms of 4 numbers, the first three being the relative strengths of the field in the X, Y, and Z axes of the instrument, and the 4th being the absolute strength of the local field.

Put yet a third way, it tells you what % of the magnetic field is along your line of site, what % is left to right, and what % is up or down. It also tells you what the absolute field strength is in some obtuse units like milligaus or microtesla. Note that the units of magnetic field strength are unrivaled in obscurity. Few engineers really have a "feel" for what a nanotesla is. All of us have a feel for what a pound of force is like, or a degree of temperature (Fahrenheit or Celsius). Scientists now believe that some migrating birds have a magnetometric organ in their brain. Possibly these birds feel that a few thousand nanotesla is a pretty nice thing, sort of like a 75°F day for us.

Micro Space Craft

Lacking such an organ, we humans must bear the burden of listening to a description of how a flux gate magnetometer works. Using an internally generated magnetic field created by flowing current through a coil of wire, it attempts to magnetize a piece of soft iron inside it along one axis. Then it immediately attempts to magnetize that piece of iron again, along that same axis, but flipping the north and south poles. It does this constantly, back and forth, back and forth. The magnetometer is calibrated to know how long it should take to flip the magnetization back and forth. Any variation in the rate that the iron sample actually gets magnetized, compared with the calibrated rate, is the result of an external magnetic field, which is exactly what it is trying to measure, reinforcing or partly cancelling the locally applied field. That is, if a component of the external field is helping to magnetize the sample in one direction, it is co-magnetized faster in that way and slower in the opposite way.

To measure the magnetic field strength along all three axes (X, Y, and Z or alternately front/back, left/right, and up/down), a magnetometer requires three flux gates, one on each axis. The magnetometers in minivans, for example, have only two gates, since minivan drivers tend to not care how much of the magnetic field is vertical. They want to know just the horizontal parts, from which you can determine, knowing that in fact the "interfering" magnetic field points north, what direction on the compass your minivan is pointed.

As complicated as this might sound, the end device is simple and easy to use. As an example, the Schonstedt Instruments SAM-73C 3 axis magnetometer is a small box, about 6"x2"x2" with a single 9-pin connector on one end. You attach a single voltage supply on one pin, and three other pins output a signal directly proportional to the magnetic field measured along the three-instrument axis. Hook these three pins up to the nearest analog to digital converter and your spacecraft computer now knows the direction and magnitude of its local field. That is all there is to it, except for all the important little details that make building satellites fun. Some companies are beginning to produce highly miniaturized magnetometers that may prove ideal for small satellites. Ithaco makes a lovely little two-axis magnetometer that features very low power consumption; unfortu-

Chapter 6—Magnetic Attractions

nately, its price tag might discourage small satellite builders on a limited budget. Three of these would make a fully redundant three-axis system.

Navigating by Magnetic Compass

Creating a mathematical model of the earth's magnetic field is the raison d'etre for satellites that go out and dutifully measure that field every day. If we know where the satellite is in its orbit, and the mathematical model tells us which way the field, at that point, is oriented, we can determine which way the satellite itself is pointing. What we don't know is the attitude about the field line. A terrestrial example: Knowing which way is up tells you nothing about which way is east.

Conversely, if we know what the satellite's attitude is, perhaps by using a sun sensor and an earth sensor, we can determine where the satellite is in its orbit by knowing what the magnetic field locally feels like. If you aced Kalman filters in grad school, you might come up with a way to figure BOTH things out, by relying on the ever-changing nature of the field to gain more information over time. But if you aced Kalman filters in grad school, you probably are only reading this article to find little inaccuracies so you can "gotcha" the author. A mind, you know, is a terrible thing to waste.

The basic principle is evident in every ancient ship's compass, even though at this point we're beyond what Columbus, Magellan, and Lewis & Clark knew about magnetic navigation. They had just needle compasses, little bar magnets that line up with the earth's field. And they didn't have to worry about the third (vertical) axis, glued, as they were, to the surface of the earth. But LEO satellites' second use of the earth's magnetic field is unique to satellites.

A magnet, subjected to a local magnetic field, feels a twisting force or "torque" applied to it if the local field and the magnet are not aligned. Sounds complicated, but most of us have seen this principle at work. Find a small bar magnet, or magnetize a pin. Float it on a piece of rubber or cork in a bowl of water. The magnet turns until it is aligned with the earth's field. This torque caused by the skew between a magnet's field and the field in which it is immersed is what forces the needles of scouts' compasses to point north.

In a satellite, we can make an electromagnet by installing a coil of wire and passing a current through the wire, we have an electromagnet. If this electromagnet's field is not aligned with the earth's field, a torque is applied on the coil. Since the coil is attached to the satellite, the torque is applied to the whole satellite, trying to turn the coil into alignment with the earth's field.

Now we have a way to cause the satellite to change its attitude, essentially to steer it, without using rocket engines and propellants—just magnetism and electricity.

But with only one coil acting in one direction, we can only make one torque. We can play four more games to create a bigger variety of directions in to turn the satellite. First, just reverse the current flow in the coil, which flips its magnetic field. The direction it tries to turn to align itself with earth's field is also reversed. Second, give it three coils instead of just one, pointing left/right, front/back, and up/down—the three so-called principle axes. Third, add two directions of magnetization to each of the three coils, which gives six different torques that the satellite can create. You end up with a plus and a minus torque (or you can think of it as clockwise and counterclockwise torque) on each of three axes.

Trouble is, if the earth's field happens to be lined up with a coil, it's not going to create any torque, because torque results from the misalignment of the earth's and the coil's magnetic fields. No misalignment = no torque. Time for Trick #4, which is to wait. As the LEO satellite moves in its orbit, the earth's field appears to vary its direction. Near the poles it is more vertical, near the equator more horizontal, and the variations away from the perfect dipole model also create many opportunities to find fluctuations in the earth's field's orientation. Thus, if the satellite cannot create the torque it wants, but can wait typically 10 or 15 minutes, that torque becomes available.

Chapter 6—Magnetic Attractions

Can You Say "Torque Coils"?

Rather than go to all the trouble to wind coils, build the circuitry to provide them with current, and control switching them on and off, you could create a torque with just a simple bar magnet. The trouble is, you can't shut it off! Some satellites have been flown with a single bar magnet as their attitude control system. The satellites point constantly with the bar magnet aligned with earth's field, causing the satellite to flip every time it goes over the north or south pole where the field goes vertical. Still, for most of the orbit, the satellite's magnet is about parallel to the earth's surface, since that is the orientation of the field when you are far from the poles.

For most applications, an electromagnet made of a coil of wire is the way to get torque. By varying the strength and/or duration of current flow in the wire, we can precisely meter out the amount of torque produced. These coils are, not surprisingly, known to satellite cognoscenti, as "torque coils."

There are two kinds of torque coils. The simplest is just a coil of wire wrapped around an insulating spindle. The budget-limited satellite builder can use a simple coil wound on a lathe, which makes a perfectly effective torque coil. You want to use as much wire and wind as big a coil as your mass and volume budgets allow, since a big coil with lots of windings gives the most magnetic gazorch for the minimum electric power. Be sure to choose a wire gauge that gives you the right resistance for the supply voltage you want to work with. Six coils like the one shown in the figure below have been steering the ALEXIS satellite for almost 2 years on orbit.

A coil with a slug of iron in it produces more magnet moment for a given amount of electric current than one without the slug. These iron core coils are standard for bigger LEO satellites, and are also found on small ones if space constraints don't allow packaging the larger air core coils.

Ithaco is the major supplier of iron core magnetic coils for attitude control. These are very linear, efficient devices, but a bit expensive. An air core coil uses about three times more power than an iron

Micro Space Craft

Torque coils controlling ALEXIS satellite's attitude and spin rate.

core for the same mass and characteristic dimension. But for small satellites, the power draw of the coils is seldom an issue. An overriding issue with very small, light satellites is residual magnetism when the coil is shut off, since very little torque is still enough to disturb the attitude of a very light satellite. Iron cores become slightly magnetized in use, and care must be taken to ensure that magnetization is fully canceled. This is not an issue with air core coils. They have no residual torque because they have no core, and hence no core to become magnetized

Magnets Chasing Their Own Tails

There is an interesting strategy to actually using any of these coils. To determine which coil or blend of coils to use and in which polarity to get the torque you want, you need to know the direction of the local magnetic field. A magnetometer is carried on board to measure that field. But turning on the coil produces a new magnetic field that is locally much stronger than the earth's field, and the magnetometer senses this locally generated field instead of the

Chapter 6—Magnetic Attractions

external field it is working against to produce the torque. If things work out right, you can get a nice unstable attitude control system out of this that is sure to provide hours of amusement for you and your customer.

Several strategies are available for turning on the coil. One is to simply switch it directly to the supply voltage rails, so that there are three possible states: ON+, ON- and OFF. This strategy is the easiest, but tends to be less than power efficient, because the power to a coil is proportional to the square of the resultant moment. It also limits the directional control of the moment. Another possibility is to control the coils linearly, which allows complete control of the resultant moment's direction and magnitude. While some interesting schemes can be used to provide a linear drive current very efficiently, AeroAstro provides people with clever little coil drivers so that they can just think about designing a satellite and not worry about how to drive inductors.

The FORCE: It Comes in Colors

Besides controlling your satellite according your own field instead of the earth's, you need to remember a few important caveats to protect you from magnetism's darker side. Magnetism doesn't discriminate. A permanent magnet or a field produced by torque coils both create torques when interacting with earth's field. But those forces are produced not just from the permanent or electromagnet you put on board for that purpose. If you have any ferromagnetic materials in your satellite (maybe those big Kovar power transistor packages) that become magnetized, they will constantly interact with earth's

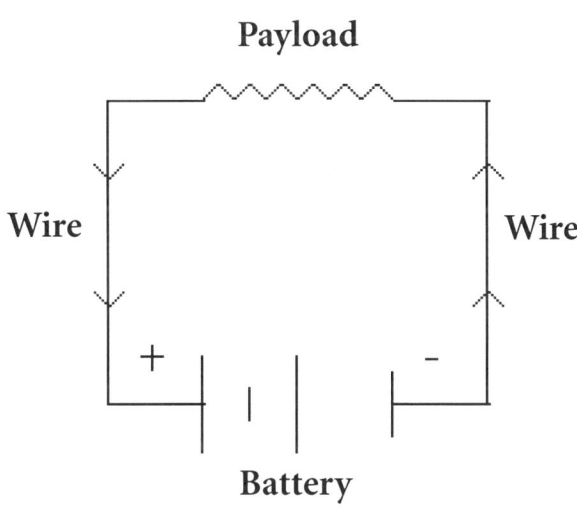

Current loops create magnetic fields.

field, steering your satellite around for you, and probably not the way you wanted it to go. Any loop of wire carrying current is an electromagnet. So if you run the plus wire out to a big power consumer, like your satellite's payload, and then bring the ground or minus wire back via a different route, the current loop you produce creates a field that will cause undesired torquing of the satellite.

Fortunately, these "noise" torques are directly correctable by the control coils. But they still must be considered and minimized because canceling them wastes electric power. More importantly, your satellite is constantly being seesawed in the tug-o-war between the noise torques and the coils snapping on and off to correct them.

The tendency of electrically charged particles to travel along magnetic field lines is a great thing if you want to build a Tokomac fusion reactor or a magnetic bottle to hold antiprotons. You can file that with your other hints from Heloise for when special guests stop by. When you fly your satellite in LEO and enjoy that nice strong field to navigate and steer by, you are like a swimmer in the Gulf Stream enjoying its nice warm water. You have, in the latter case, at no additional cost, the opportunity to enjoy the company of Portugese Men o' War (so far as I know these have not yet been renamed Persons o' War), jellyfish, stingrays, and the odd shark seeking refuge from cruising the chilling North Atlantic waters in search of a seal or downed naval officer. In the case of the earth's magnetic "stream," your companions are all kinds of charged particles that are wafted our way from the sun in what is known as the solar wind.

Atomic nuclei, electrons, and atoms missing a few of their electrons, causing them to be positively charged, really like our magnetic field. These tiny particles are "energetic," meaning they are moving real fast. They pass through electronics enclosures and either wedge themselves in your integrated circuit substrates, or zip right through them. In either case, they can deposit a bit of charge as they go, which has a nasty tendency to make 0s out of your 1s and 1s out of your 0s, which is disturbing, and can be fatal, to your on-board computing systems. These charged-up speed demons can damage materials just by passing through them—sort of like putting a tie

Chapter 6—Magnetic Attractions

tac through nearly the same spot on your favorite tie, the one that everyone in the office is totally sick of, every day. Eventually the office will get its relief!

Charged particles tend to be concentrated in the three places where the field lines themselves are concentrated: the magnetic North Pole, the magnetic South Pole, and one place off the coast of Brazil, aptly and ominously named the South Atlantic Anomaly. Satellite electronics designers spend a lot of time on this issue. We look for components that resist damage from these particles, and integrated circuit materials that do not tend to absorb charge from passing ions. These exist, but Murphy's Law prevents the component you really want to use, the one that would simplify the whole design and make your life at work really worth living, from being available in those materials. Instead, you end up designing around some power-hogging artifact of a previous decade just because it's pretty impervious to atomic guerrilla warfare. Alternatively, circuits can be buried inside of metal enclosures, thick ones, usually made of dense stuff like lead and tantalum. But at $1000 per pound launch costs, this solution tends to not create harmony with your program manager and customer.

For More Info:

Following are a few references on some space applications of magnetics. The basics of magnetism are covered in virtually all collegiate introductory physics textbooks.

Spacecraft Attitude Determination and Control, James Wertz, Ed., Kluwer Academic Publishers div. D. Reidell Publishing Co., Dordrecht, The Netherlands, 1978.

Psiaki, M., F. Martel, and P. Pal, "Three-Axis Attitude Determination via Kalman Filtering of Magnetometer Data," *Journal of Guidance, Control and Dynamics,* Vol. 13, No. 3, May-June 1990, pp. 506-14.

Psiaki, M. and F. Martel, "Autonomous Magnetic Navigation for Earth Orbiting Spacecraft," *Proceedings, Third Annual AIAA/USU Conference on Small Satellites,* Logan, Utah, Sept. 1989.

Chapter 7
Everything You Always Wanted to Know About Radio:
Part I: Shatter the Myth of the Digital Miracle?

L ET'S! Does it irk you that your electric service account number has enough digits to specify the date of birth, sex, hair color, height, and weight of every human on earth? Does anyone really believe that before software all R&D programs came in on time and on budget? A 1 µs dropout of the AC power mains means a million people are late for work, and 100,000 person hours are lost resetting clocks flashing 12:00 AM. Ever lose 100 pages of your handwritten work because the phone rang at a critical moment? How much leisure time has AUTOEXEC.BAT given back to you?

Digitalization has subtler costs. Anxious to join the herd plunging off the edge of the digital cliff, engineers have deserted everything mechanical and analog. This is obvious just looking around our post-digital but otherwise Neanderthal world. You have a digital thermometer to tell you precisely how poorly your office temperature is regulated. Your all-digital car, which still has a spare tire in the trunk the size of a 1985 PC Portable, is still made the way they made 'em in 1937—sheet steel, reciprocating 6 cylinders, glass windshields and pneumatic tires. The digital "information center" is always at the ready to tell you that "Fuel Is Low" so that you don't abruptly stop burning one of earth's most precious, and noxious—when mined, shipped, refined, and burned— resources. The 737 is the loudest, least comfortable passenger jet airplane ever made, but its flight displays are gorgeous!

Micro Space Craft

Radio engineering, once the leading edge of electrical engineering, has fallen into disfavor with the digital fiber optic crowd. Radios—eeeuw, aren't they analog? (Horrors!) Anyway, everybody knows how radios work. We're practically born with radios and televisions glued to our ears and fixed in our line of sight. Most of us spoiled children of the second half of the twentieth century were conceived with either a radio or a television on somewhere in the room. As infants lying in cribs our breathing is transmitted by little Fisher-Price radios all over the house, a legacy of potentially tremendous value to aliens listening to earth's strange radiations! We don't have air engineers to study breathing, or water engineers to study drinking. What's the big hassle with radio engineering? Insert 12 D cells, heft that blaster onto the left shoulder, and boogie. Don't bug me about radios right now. I'm on the cellular phone. Radios have become so embedded that, like your car's engine, unless you own an MG or a '65 Ford Fairlane with 289 V-8, we really don't notice them or need to understand them any more.

Satellites have a few pesky qualities about them that make their dependence on radio rather significant. For one thing, they are far away. It is very hard to make wires 400 or 40,000 km long, hang them from space, and not get them all tangled up. Plus, the visual pollution is unacceptable to EPA. They move fast (satellites, not EPA)—7 km/s (15,000 miles an hour), which makes reeling out cable a bit clumsy anyway. Yet they are pretty useless if we can't exchange information with them. All satellites, whether they are relaying TV, radio and telephone signals, measuring the height of the ocean, tracking a volcanic plume, or watching for Gamma Ray bursts, are information machines. Radios move that information down to us.

I loved smoke signals as a kid. I was your normal eight-year-old pyrotechnic. Smoke fascinated me, mainly because where there's smoke, there's the possibility of burning down your parents' house, including the arrival of neat fire engines, water spraying all over the place and the promise of a dramatic rescue of the dog like the Dad on TV did. That's how little boys think, which kind of motivates one to rethink this whole latchkey child thing.

Chapter 7—Radio: Shatter the Myth?

But smoke signals are radios. Native Americans didn't send the actual puffs of smoke to neighboring tribes to signal a shortage of water or whatever. The smoke modulated the transmission of radiation, sunlight. The receiver, a friend at the next settlement with eyes to detect sunlight, picked up this modulation, which consisted of dark patches of smoke interposed with clear areas. Modulated radiation = radio. Not a perfect definition, but it'll get us where we need to go.

Morse Code is a way to represent all the letters of the Roman alphabet, plus numbers and punctuation marks, with dots and dashes. "A" is dot dash (• –), "B" is dash dot dot dot (– • • •) and "Hey, I gotta get the heck outta the radio room or I'll be late for the kids' soccer practice. See you later, dude, say hey to Sarah and the kids" is

– – • • • • • • – –

Abbreviation was elevated to high art by the early radio operators.

Sending the long and short pulses of Morse Code by radio is not so different from sending smoke signals. You have a transmitter that you can turn on and off, and a receiver sensitive to what you are transmitting. The simplest transmitter may be a light with an aperture you can open and close. This was briefly popularized by Paul Revere—not the '60s rock band or the copper saucepans, the 18th century revolutionary who earned his living when not warning of Redcoats by making copperware. In fact, much of 19th century Europe was linked by optical telegraphy stations operating in a giant international network. It is still used today by ships at sea. SOS, aka Save Our Ship, is just

• • • – – – • • •

a relatively easy pattern to remember, even while some of the North Atlantic flows over the aft deck. Flashlights are good Morse Code senders.

But light isn't radio. It's light, right? Yes, it's light, but light is just one type of electromagnetic radiation, a type we sense with our eyes. Microwaves reheating last night's tuna casserole are another form of that radiation, as are x-rays that form images on photographic film used to look at our bones. Infrared radiation, the way you can "feel" from a distance that a turned off stove top burner is still warm, is also electromagnetic radiation.

When electricity flows in a wire, it creates a field around the wire called an electromagnetic field. This idea sounds very scientific and mysterious, and I still think it actually is, but it is easy to visualize nonetheless. A current flowing through a wire creates alignments in the atomic nuclei and electrons that form the wire material, allowing their electrical and magnetic properties, which are normally all random and cancel each other out, to align themselves and cooperate. One result is that a compass brought close to a current-carrying wire will point in the same direction as the wire. Electromagnets are just coils of wire carrying lots of current to make a very strong alignment of their magnetic dipoles. Very handy for picking up black Lincoln Continentals containing bad guys in James Bond films and dropping them unceremoniously into crushers—another fantasy common in 8-year-old boys.

If the current is steady, or varies only slowly, the fields are steady and about all we can do is make magnets and heaters. The "heads" of a cassette tape recorder or a floppy disc drive basically work this way. These wires with current flowing in them make magnetic fields that "write" on the magnetic materials inside the tape or disc.

But if we make the electric field fluctuate, that is, flow first one way, then the other, the magnetic field propagates, or travels. Why? Because some energy is required to reverse the flow, and some of this energy escapes to propagate itself away from the source. At very low frequencies very few reversals of flow occur, so this transmission is not too efficient. The current that flows in our electrical wiring at home fluctuates, that is, reverses direction and then reverses again to the original direction, only 60 times a second, which is considered VERY slow indeed. If you drive under a high tension (high voltage, but also high current, which is the key to strong radio energy transmission) cable with your car's AM radio on, you can hear

Chapter 7—Radio: Shatter the Myth?

the 60 Hz (Hertz = cycles per second) hum. Originally popularized in horror movies of the '40s when they were getting ready to electrocute some ghoul (yet another fascination of eight-year-olds), the 60 Hz hum has become, in our society, synonymous with electric power.

For some reason, frequency waves as low as 100 or 200 Hz propagate well through water. Used for transmitting radio waves to submarines, they are otherwise pretty useless. Aside from requiring huge antennas to resonate efficiently with their very long wavelengths, they are bad information carriers. The human voice can fluctuate at up to 8000 Hz, but a 100 Hz radio wave cannot vary any faster than 100 Hz, and practically cannot be modulated faster than 50 Hz. Carrying a human voice is practically out of the question, unless we want to transmit the voice very slowly and then play it back later at the right speed. In which case, a 10-second burst of speech could take 15 minutes to send. Thus, 100 Hz class radio waves, with very limited application, are called very low frequency (VLF) for good reason.

Radio waves really don't get interesting 'till you get to high frequency (HF). HF happens around 100,000 Hz, or 100 kHz, a value known as "medium wave" that you can find on AM radios in Europe and Asia. Our American AM radios start at 550 kHz. These frequencies have two significant advantages. The waves can be efficiently transmitted and received using more compact gear, and they vary fast enough that we can modulate them with human voice and other sounds that our ears can hear. The highest frequency, or rate of variation, that we can hear is about 20,000 Hz. A radio wave at 100,000 Hz varies five times faster, so it has plenty of ability to vary fast enough to carry, or be modulated by, all the sounds we can hear.

Voilà! Radio. For a long time, people were pretty excited to radiate at 100 kHz or so. After all, it isn't easy to make something vary at 100 kHz. The fastest your car engine can spin is maybe 50 or 100 Hz (0.1 kHz), so these radio transmitters, were they mechanical devices, would be really whizzing! But after a while…

Micro Space Craft

The most common way to make electricity cycle is with a piezoelectric crystal. These quartz crystals are not unlike the quartz you find in nature as a clearish rock, or as a component of granite, which we all are told in elementary school is quartz and feldspar, even though we have no idea, nor could we care less, what quartz or feldspar are. God's little gift to radio is that when you squeeze a carefully cut sliver of quartz, a bit of electric current comes out, and when you do the opposite, put a current into it, the crystal contracts or expands a little. Really a very neat thing. Modern micromachines use this property of quartz to affect very small motions, like for aligning precise optics or soldering wires into miniature electronic circuits.

In radio applications, we apply a brief spike of current to the crystal, and it expands and then relaxes and vibrates in that mode for a while, like a bell. As it vibrates, we get a vibrating electric current. These vibrations can go up to 100 million vibrations per second for very fine slivers of crystal.

In the early radio days, making such fine crystals wasn't possible. The pioneers used bigger hunks of crystals, and got resonances in the range of 100 kHz to 2000 kHz, two million Hertz, or two Megahertz (MHz). The standard AM radio band that extends from 550 kHz to 1.6 MHz was allocated because it was the highest frequency range of radio signals that could be practically produced and detected by the modern radios of the 1920s. Today we more typically use television and radio frequencies of 80 MHz to 160 MHz. The pioneers thought they were really sizzling to get 2 MHz, so they subsequently called the range 3 MHz to 30 MHz "High Frequency." We call frequencies of 30 to 300 MHz "Very High Frequency" or "VHF."

The press to get to higher and higher frequencies was motivated by several factors. One was space. When all the frequency space is used in a particular band, in other words, when transmitters broadcast on all the spaces between 550 kHz and 1.6 MHz, your only choice is to go a little higher. By the end of WWII, practical radios had been built with frequencies over 1 GHz. That's one billion cycles per second! Today, without going to any technological extremes, we can produce radio energy with frequencies as high as 100 GHz, or 100 billion cycles per second.

Chapter 7—Radio: Shatter the Myth?

The range from about 300 MHz to about 3000 MHz (3 GHz) is called Ultra High Frequency (UHF). If you have ever shopped for olives, you know where we're headed. The smallest olives you can buy are Large. Probably 75 years ago olive growers were pretty impressed when they grew a few of those Large guys. But agriculture, like radios, moved on, bringing to our grocery store shelves the Giants, which are still easily mistaken for a bloated raisin. Grape growers were no way going to be outdone by the olive mavens. Then you get into Mammoth, which are nowadays the size you find in the lonely little salad bars cast into the corners of rural grocery stores as a gesture that says, "We know city people eat this stuff, but Lord knows why." Big olives, the kind you want to serve when the boss comes over for dinner, are known as Colossal and Super Colossal.

Why should electromagnetic waves be any different than olives? Today the rage is Super High Frequency (SHF). Somehow I always expect these waves to come flying through the window wearing blue tights and a red cape.

Enough radio talk. You probably have a few disks that need to be backed up, mail to read off the Internet, or maybe your answering machine's digital outgoing message needs refreshing and you need to get to the store and pick up some new software for the kids. No problem at all! Take a couple months' break. Radios have been around 100 years. They'll wait. No rush to find out what S, L, and X have in common besides the word "Suzuki"; why anybody really needs to know the speed of light; and whether a crack team of government proposal reviewers would ever put grant money into radio.

Chapter 8
Everything You Always Wanted to Know About Radio:
Part II: Faster Than a Speeding Bullet

NO PRIMER ON RADIOS is complete without the graphic below, showing the progression of radio waves from VLF power transmission, through the HF, VHF, UHF spectra, on up into microwaves, up through SHF, then into infrared and visible light, ultra violet, X-rays and Gamma rays. Wherever you roam in the spectrum, the physics is the same—alternating electromagnetic fields.

All this electromagnetic radiation propagates at about the same speed. What junior scientist didn't learn that light travels 186,000 miles (300,000 km) EACH SECOND. Fast! Bullets get up to maybe 1 mile a second, and sound only travels 1/5 of a mile in a second in air. 747s and other jets travel more like 1/7 of a mile per second, less than one millionth as fast as light. Even our satellites whizzing around at 4 miles a second are less than 1/40,000th the speed of radio and light waves.

Given that all electromagnetic radiation travels at about 300,000,000 meters a second, we can define the wavelength as the distance the radio wave travels in 1 cycle. For a signal on your AM radio at 600 kHz, the wavelength is given by:

wavelength, λ = 300,000,000 m / s ÷ 600,000 cycles / s = 500 meters (about 1650 feet).

Micro Space Craft

Note that fancy Greek lambda, λ, which is the symbol engineers use for wavelength. We also use another classy letter, Ω, omega, for frequency. So a wavelength of a signal in the AM radio band is somewhere around 500 meters long, maybe only 187.5 meters (615 feet) at the top of the band around 1600 kHz.

It's hard to carry an antenna 200 meters long around with you when jogging. If you look inside a portable AM radio, you'll see a wire almost that long tightly coiled up. That's your antenna. FM radio (88 MHz to 108 MHz) wavelength is 3.4 to 2.8 meters. Your car's antenna and the headphone cord from your walkman to your ear are both about half that length, making them pretty good antennas. Cellular phones work at 850 MHz, where the wavelength is 0.35 meters or about 1 foot. A "half wave" antenna is the size of those little black curlicue things you see sticking up from the rear window of almost every car in the world with a sticker price over $25,000. Efficient antennas for cellular telephones.

One little quirk of satellite radio nomenclature: most radio people outside the satellite world talk in terms of bands—ranges of frequencies or more often of wavelength. The latter is preferred since it's the wavelength we have to deal with most, like what length antenna, what size features on the circuit board, and so on. Not satellite jocks. They talk about L band, S band, X band. Lucky you, I'm too young to tell you all about how they made the first microwave radios.

Note here "microwave" refers to frequencies ranging from the upper UHF into SHF. In fact microwave ovens run around 2.5 GHz, 2,500 MHz, or a wavelength of about 0.1 meter or 4 inches. That happens to be a resonant frequency for water molecules, which are efficient antennas for that wavelength, and that's how microwave radiation cooks food. Unfortunately, cats are made of mostly water, which has generated a lot of misdirected speculation, epicentered in the eight-year-old-boy segment of our population, on the potential for inserting cats into microwave ovens. I like to think nobody has ever actually zapped a cat, and that it's just so much macho playground talk.

With all the excitement in recent years about the possible health effects of radio waves, you might be wondering about whether the

Chapter 8—Radio: Faster Than a Bullet

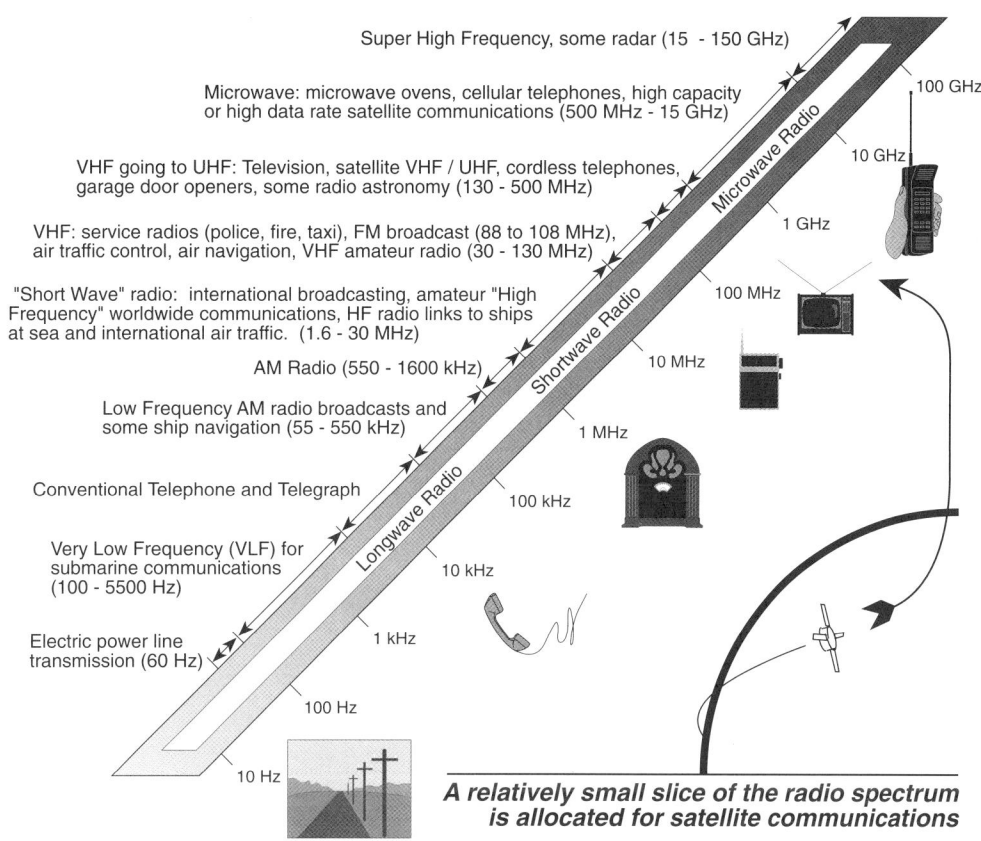

A relatively small slice of the radio spectrum is allocated for satellite communications

disappearance of radio engineers is caused by other reasons than disinterest and fascination with bits and bytes. Certainly, the radiation of a microwave oven is dangerous above a certain level. Above that level, the cells in your body are heated and could be killed. Not good, and no controversy. This explains why, for instance, microwave ovens are carefully built so that the door has to be closed before they turn on. Despite initial controversy, I would suppose microwave ovens are a lot safer than the stoves and conventional ovens we all take for granted—no grease fires, no burned fingers, no gas leaks, just to name a few hazards that can accompany conventional cooking. I doubt a microwave oven has ever caused a house to burn down! The arterial effects of buttered microwave popcorn are, however, another possible consideration.

But as you move away from the wavelengths that resonate with biological materials, which are mostly water and hence mostly 2500 GHz, and thus eliminate heating effects, the health picture is much less menacing. Standards have been set for maximum exposure to radio energy, and engineers who work around radios, particularly microwave radios, where the standards dictate the most stringent controls, need to be conscious of this limitation. But cordless and cellular phones, electric power transmission lines, desktop computers, and induction stoves radiate at wavelengths very far from those needed to couple to the molecules in our bodies, and their radiation is weak. Handheld cellular phones have 0.8 watts of transmit power, about the same as one of those little key chain flashlights. No clear evidence indicates that radio waves from these sources have any health effects. One study did show increased incidence of disease in people living near power lines, but it turned out that the power company used vegetation-killing chemicals under the lines to keep trees and bushes from shorting to them, and these chemicals were dangerous to human health.

It is virtually certain that safe, normal usage of radios has no effect on your life expectancy or health, but people will continue to research radio energy's biological effects for many years. While no clear danger inherent to our current handling and application of radios exists, possibly some of our uses of radio and power transmission may have statistically significant effects on health over large human populations. Current research on radio safety focuses on this subject.

Back to the radios themselves and the arcane nomenclature of the radio bands used to talk with satellites. Because resonant microwave cavities were created with various oddly shaped devices, early microwave engineers named the wavelength bands after their resonators. For instance, L band is around 1000 MHz (1 GHz). S Band is 2 GHz. X Band is 8 to 10 GHz.

You might already guess three reasons satellites like these very high frequencies. First, you need a high frequency carrier to send a lot of information fast, like a stream of images from a telescope, or to relay television. Second, weight and space are at a premium on

Chapter 8—Radio: Faster Than a Bullet

satellites. Microwaves are short, so their antennas are much more compact. (Otherwise we'd call them macrowaves.) Finally, by the time people invented satellites, we'd found other terrestrial uses for all the frequency space up to several hundred MHz. To a technical person, this might sound a bit political. But I doubt the world was ready to trash all of its brand new television sets to make spectrum space for the first satellites. The real world is like that.

Just to keep us guessing that someone might be looking out for us after all, it turns out very handy that people were so attached to the sub-100 MHz spectrum. The highest levels of earth's atmosphere are busy shielding us from nasty ultraviolet, gamma, and X-radiation. In doing so, these layers become reflective to radio waves up to about 50 MHz. In the days before satellites, this feature of the ionosphere was the only means we had for transoceanic communication by radio. High Frequency (HF) radios transmitted East from the coast of Virginia, eventually bounced off the ionosphere, and were received in England. Lots of services still use this means of communication, popularly known as shortwave radio. Most of the world's countries broadcast their cultural and political images daily on the wavelength bands between about 6 MHz (50 meters) and 21 MHz (15 meters). Ham radio operators using low-wattage transmitters equivalent to the brightness of a flashlight bulb communicate all over the world bouncing signals off the charged ionosphere.

So there would be a couple of problems with using HF radio and ionospheric reflection for what satellites do—relay telephone and television and bring data down from space. For one thing, the frequency is low, so the data rate you can get is limited. A few television channels would fill up the entire usable wavelength band. Forget 200 channels of satellite TV. Ditto telephone calls, or plan on long queues to get a turn to use the available throughput capacity.

But what's maybe even more troublesome is that propagation of HF radio signals by means of ionospheric reflection varies randomly from minute to minute, and it varies strongly as the reflecting area moves from sunlight to darkness, also with the seasons, and even with the sunspot cycle. Thus, HF links are undependable and often noisy. HF radio is also unsuitable for satellite communications because ionospheric reflection strongly attenuates the signal along its path to the satellite.

For all these reasons, satellite communication has focused on the VHF, UHF, microwave, and SHF bands, that is, everything above 30 MHz. Terrestrial long distance communications rely on the HF bands from about 3 MHz to about 30 MHz. In fact, most satellite operations begin above 100 MHz. What's left between 30 MHz and 100 MHz? Police radios, aircraft radio, remote controls, cordless telephones, FM radio, and a bit of television—stuff that we do NOT want to propagate over long distances. Even so, many services are abandoning the lower frequencies for higher ones because their electronics become smaller and their antennas more compact.

What's been described so far is like drawing the geopolitical globe of the world. We've established:

- the locations of the oceans and the continents, that is, the use of VLF for submarine communications and power transmission,

- "regular" frequencies for AM radio for historical reasons,

- HF for international shortwave radio using ionospheric reflection, and

- VHF, UHF, microwave, and SHF for satellites, for line-of-sight terrestrial communications including cellular telephones, FM radio, television, and cooking dinner.

Like countries vying for the finite real estate on their land masses, various radio services compete and divide up the spectrum. Telephone and TV transmission, police radio, scientific satellites, garage door openers, microwave ovens, even hydrogen molecules, whose vibrational modes at microwave frequencies form the basis for radio astronomy—all of these have little patches of spectrum allocated to them. The continuing distribution and redistribution of this resource helps the editors of Space News keep their pages filled,

Chapter 8—Radio: Faster Than a Bullet

employs hard working lawyers all over the world, and provides the focus of international conferences where the agreements on frequency allocations worldwide are hammered out, or not.
I'll leave the political commentary for Art Buchwald and the Maison Blanc set, and we'll stick to our photons, shall we?

Tune in next chapter for the exciting conclusion to the Satellite Radio mini-series. A torrid discussion of radio power and amplification is followed by passionate confessions about ground station links. Issues of Doppler and modulation throw a bit of spice into the story. Don't miss the surprise ending!

Chapter 9
Everything You Always Wanted to Know About Radio:
Part III: What's Up, Doc?

WHAT'S A WATT? Let's not get too technical. A typical light bulb puts out 10 or 20 watts of light. The other 80 or 90 watts in a typical 100 W incandescent light bulb are dissipated as heat. A flashlight puts out maybe two watts of light. The standard S-band (2 GHz) spacecraft transmitter, used on many low earth orbit satellites including science satellites and Landsat, the NOAA polar orbiters downlinking weather data from low earth orbit, is a two-watt transmitter, also about as much brightness as a flashlight. Why so dim? Just like a flashlight, to make two watts of radio energy at 2 GHz requires much more power, like about 15 to 20 watts, which is a significant power draw for a small or medium size satellite.

A satellite coming into view as it rises over the horizon in a 750 km (400 mile) orbit is about 2000 km (1400 miles) away. We routinely communicate across these distances with power so low that it is the equivalent of expecting a flashlight flicked on in Boston to be seen in Miami. Yet we can detect it flashing on and off thousands, even millions of times a second.

The lingo for the connection of a receiver and a transmitter is "link." Many factors go into establishing the link between a satellite and the ground. The transmitter's power spreads as it propagates from the satellite. Assuming a nondirectional, or isotropic, antenna, which is often used on a small satellite so that it doesn't have to precisely point at the ground station, the transmitted power spreads into ever increasing spherical shells. By the time the transmitted radio power has traveled 2000 km, it is 4 trillion (4,000 billion) times weaker than it was 1 meter from the satellite. Starting out at about 1 watt of radio power for each square meter, the power density on the ground is less than one millionth of a millionth of a watt per square meter.

Yes, there is a lot of technology behind receiving that infinitesimal signal. But we should not forget that it is pretty amazing. It's a miracle we take for granted every day. If radio didn't exist and you proposed to ARPA or NASA to put a two-watt transmitter thousands of miles away and receive it perfectly with a relatively simple set of equipment, your proposal would certainly be dismissed along with the perpetual motion machines and momentum drive rockets. Luckily the radio pioneers, like their aircraft-inventing counterparts, came along before the government got into funding its development!

Besides Divine Providence, we do rely on a some ingenuity to intercept these radio signals. The most visible feature of satellite ground stations, their dish antennas, create a big area for intercepting satellite-transmitted radio energy and focus the signals being broadcast to the satellite. A six-foot (about two-meter) diameter dish has a beam about 6° wide—an amplification factor of about 1000 compared with a non-directional antenna. Ignoring higher order effects, it's interesting that this focusing effect is frequency independent. A six-foot dish could just as well receive light energy from a two-watt light bulb as a 2 GHz RF signal. We call such an "optical" dish by another name, a six-foot telescope. Given the fine surface required to focus the shorter wavelength light, a six-foot telescope is just a bit more expensive than a typical small dish for a satellite ground station.

Having collected all the signal it can, the receiver's efforts are focused on amplifying, or increasing the signal strength by a factor of way over a million times. The S-band receiver flown on the ALEXIS satellite, which is not atypically sensitive, had an amplification factor of one trillion, meaning that it multiplied the radio signal coming into it by a factor of a million times a million. The fundamental problem with extreme amplification is that both noise

Chapter 9—Radio: What's Up, Doc?

and signal are amplified. Thus, the receiver must be selective as well as sensitive enough to reject signals near to the signal of interest. Even with the lowest noise amplifiers and the most selective filters, the signal remains below the power level of the amplified noise. Special detection schemes are employed that "vote" on whether perturbations in the noise indicate a signal or just random noise fluctuations. Encoding schemes increase the amount of raw information sent so that when an error in decoding the received signal occurs, it can be detected and fixed, or the satellite can be requested to retransmit a part of its message.

While extremely low signal level detection is virtually unique to satellite links, even the simplest AM/FM radio employs many of these same techniques, including highly selective tuners and very high gain amplifiers to multiply signal strength typically by a factor of a million or more so than you can tune in NPR or MTV. The large distance and high data rate coupled with the power handicap of transmitting from a satellite compound a problem that has always been a technological challenge.

Unique to Low Earth Orbit satellite communications is correction for Doppler, which is the apparent frequency shift caused by relative motion of the transmitter and the receiver. As the satellite first approaches and then recedes from the ground station during its brief overhead pass, the satellite transmitting frequency, as detected on the ground, appears to rise and then fall, just as an approaching car's horn, a train's whistle or a jet airplane's roar seems higher pitched then lower pitched as the moving object approaches, passes, and moves away. Those examples are such clichés that you might think they're the only times you experience Doppler. They're not. How many waves does a body surfer cross each minute? Several, when the surfer is swimming out from the beach, versus basically zero as the surfer coasts in with the waves. That's Doppler. The "red shift" that astronomers argue is evidence of the big bang is the shift of wavelength of light coming from astronomical sources receding from earth. Doppler.

A ground station transmitter appears to shift upward and then downward in frequency as its satellite passes overhead. The simplest way to maintain the link during this rapid variation in frequency is to have an automatic frequency controller on the ground and flight receivers constantly adapting to the changing carrier frequency. But with very weak signals, this control is difficult to implement reliably. Rather than build this complexity into the satellite, some systems include compensation for Doppler shift in the ground station design.

Ground-based Doppler is not painless either, but aside from where you put the complexity, there is a physical reason to compensate for Doppler, and that is reducing the time wasted tuning around randomly across the range of frequencies Doppler could give you until you find the signal you're looking for.

Assuming your link is not way overpowered, the time it takes to acquire a signal is inversely proportional to its bandwidth. At 9600 bps, it's not a big deal. At 300 bps, sometimes used for low-power beacons from satellites, with the magnitude of Doppler you get at S-band; many minutes or even hours can be required to acquire a signal using a single channel receiver. In the time that a LEO satellite is typically in view of the ground station, about 5 to 10 minutes, no information can be transmitted, if all the time is spent trying to compensate for the Doppler shift!

Thus, LEO satellite ground station systems are often designed to calculate, by knowing the orbit of the satellite, second by second, what the amount of the Doppler shift is, then change their transmit and receive frequencies steadily so that the satellite senses an unchanging uplink frequency despite the Doppler variation.

Another tool to aid detection is the type of modulation used. Probably the simplest modulation is the on/off system used in smoke signals and in the early telegraphs. But on/off is actually quite difficult to detect because the receiver cannot lock on to a transmitter that is off, and it takes time to re-lock each time the transmitter comes back on. Television and FM radio use frequency modulation, hence the term FM. The transmitter is on continuously, but rapidly varies its frequency very slightly according to the modulating signal, whether that is voice or the 0/1 bits of a compu-

Chapter 9—Radio: What's Up, Doc?

ter data stream. FM has similar lock problems. The frequency is constantly varying and the receiver, seeing the signal through a mask of noise, easily loses lock to the satellite. Phase Shift Keying (PSK) is the most effective, practical modulation scheme. Neither the frequency nor the amplitude changes, ever, allowing the ground receiver to establish and maintain lock very reliably even in low signal conditions. What changes is the phase of the transmitted wave. As shown below, the radio wave is advanced or delayed slightly with each bit of modulation, and this phase shift is detected. PSK coupled with encoding, which involves sending the same information multiple times to provide a cross-check and a means to detect errors, provides a factor of 10 to 100 in increased sensitivity to very weak signals.

Phase shift modulation is by nature digital. The phase of the signal is either shifted or it isn't. Of course, you can shift it by varying

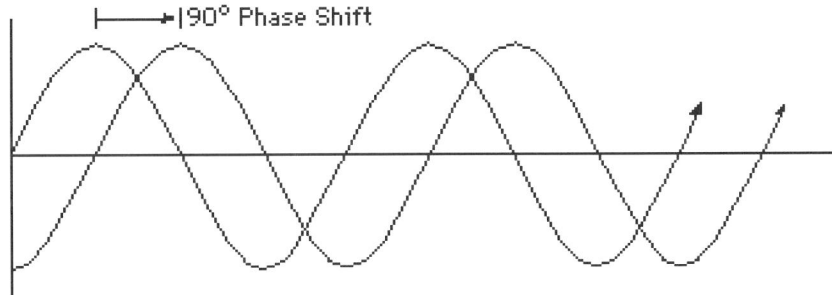

A sine wave phase shifted.

amounts, but most PSK (Phase Shift Keying) systems use either 180° (BPSK for Bi-Phase) or 90° (QPSK for quadrature phase) shifts because differentiating between small phase shifts is difficult. Phase Shift receivers are typically designed to detect the existence or non-existence of a particular phase shift angle. In BPSK, this equates directly to detection of a 1 or a 0. In QPSK, the phase shift corresponds to two bits of data instead of one. For instance, the detection code might be:

103

Micro Space Craft

PHASE SHIFT	INFORMATION
0°	0, 0
90°	0, 1
180°	1, 0
270°	1, 1
360°	0, 0 (same as 0°)

Back when digital wasn't everything (remember LP records and clocks with round faces that were called "clocks" instead of "analog clocks"?), FM was popular and a lot of work was done sending tones. Touch tone phones are a good example of this scheme. Each tone or combination of tones equates to a "message." A filter in the receiver discriminates among the tones. The system is simple and reliable, but the data rate is painfully slow, which means that sending gobs of telemetered data, what people routinely do nowadays, is impossible.

Many satellites were flown that could only detect about 10 tones. With each different tone, the satellite undertook a specific action, like turning a radio transmitter on and off. That's fine as far as it goes, but if you want to send a few color images in the form of 100 MB of data, and each tone gives you only 2 or 3 bits, you are basically left to dial 250,000,000 tones, which takes a while! Tone modulation is still used for simple systems. One of the most basic is command destruct systems on board launch vehicles, which don't have too much data to move around. Basically, it's do you command it to destroy itself or not. In this application, the simplicity of the tone system, and the fact that the rocket gets continuous reception of the ground station signal, nominally keying the "don't blow yourself up" tone, adds safety to the system.

Radio has become a very political technology. Only a finite amount of frequency bandwidth exists—the one-dimensional landscape of the electromagnetic spectrum. That landscape is constantly being divided and reallocated, a process that it seems like only lawyers have the patience to graciously endure. Ten years is not a long time to create an allocation. In fact, we technical types have partly created

Chapter 9—Radio: What's Up, Doc?

our own congestion. We like clear channels—slots reserved just for our use. In fact, we don't use them very often. A solution to this waste of geography is spread spectrum, where a signal is spread out over a large area. Another person using a different spreading pattern won't even notice you, and many hundreds or thousands of users can share spectrum without interference. The battle between the older territorial division and the newer Code Delay Multiple Access (CDMA) technologies is going on right now. The lines are drawn as you might expect them to be. The existing services that "own" frequency space like the older approach, and all the new guys are pushing CDMA.

Orbital Slots. What are they? It's easy to get the impression that there's only so much room for satellites up there. The circumference of the geosynchronous orbit track around the earth, 315,000 km long, is populated by only about 100 or so satellites. For every satellite, there is 3150 km of space, about 300,000 times bigger than the satellite itself. Each satellite appears to hang above a particular point on the earth because at geosynchronous altitude, the satellite's motion around the earth is synchronized with the earth's own rotation. (See Chapters 3 and 4.) They both go around once a day, the satellite moving in formation with the earth's surface below it.

What is tightly packed is the antenna patterns. These satellites all use the same frequency spectrum, and the ground stations uplinking to them have beams that become fairly wide way up at 40,000 km from the earth's surface. Similarly, the satellites use the same downlink frequencies. The downlink beams spread a lot too, usually more than the uplink because satellite antennas are generally smaller than ground station antennas, so their focus isn't as narrow. Satellites must be spaced wider than their beam width. It is the potential overlap of these beams that determines how many "orbital slots" are available.

Low earth orbiting satellites have a much shorter period of rotation around the earth. Typically, they complete a circuit in 100 minutes or fewer. Thus, they do not appear synchronous, and hence there are no orbital slots. By the nature of these low orbits, all satellites overlap each other's coverage areas, which is a potential interference problem. Some of the solutions to this problem include licensing the satellites in discrete areas around the world, allocating each LEO satellite system with slightly different frequency spectrum, or implementing various spectrum sharing plans where they mutually agree to accept some interference.

One potential solution to provide more spectrum space is communications by laser. The laser light is modulated just like a radio wave, either by amplitude variation (including rapidly turning the beam on and off) or potentially by frequency or phase modulation. Simple on/off coding is the primary candidate, because it is easier to turn lasers on and off very quickly than to modulate their frequency or maintain tight phase coherence. In fact, laser communication is desirable because of its huge information-carrying capacity, as much as a million simultaneous voice channels on a single beam. Lasers spread much less than radio because they intrinsically create highly focused light, the result of both the laser's light generation physics and the much shorter wavelength of light than radio. Thus, the orbital slot problem basically disappears and a very large number of satellites can be accommodated without interference.

Unfortunately, the same short wavelength that makes focusing the laser light possible also makes it almost impossible to penetrate clouds, which is a feature of radio most people never think about. After all, you expect your radio or TV to work even in cloudy weather! For laser communications to work, several ground stations are required to ensure at least one of them has a clear sky between it and the satellite.

Today's laser technologies are another issue. They require copious electricity to make a very modestly powered laser. Electricity on orbit is a precious commodity. While new technologies are being developed, laser communications in space remains only a research topic. Most of the effort is directed at satellite-to-satellite links which, since they do not face attenuation in the atmosphere, use very low power lasers.

Have you ever re-carpeted? I never paid much attention to carpet until I had to pick out a houseful of it. Colors? Sure! 10,000 colors

Chapter 9—Radio: What's Up, Doc?

and a semi-infinite number of combinations. There are also weaves, materials, stain protection, long and short pile, wall to wall, and borders. Even carpet tiles abound—little 12" squares of carpet you glue to the floor—instant carpet, plus you can make huge checkerboards and produce other digital art forms, albeit at low resolution.

If you go the tile route, several other doors present themselves for you to open, like vinyl tile, rubber tile, ceramic tile, even natural wood tile. Oak, cherry, bleached wood. Wood comes in some very interesting tiles, like long, thin ones people call "boards" and "planks." Your floor can say Manhattan or Mt. Vernon, or Miami, Milano, Maccau or Manchester or just lie there and take the abuse of daily life. Pretty soon your previously wall-to-wall suburban enclave looks like the inside of the Holden Arboretum with a few Pirelli rubber areas where people are likely to spill water or drip mud. You have graduated from carpeting to "flooring." Suddenly you see floors in all their myriad relationships and possibilities! To think my wife still considers a hardware superstore boring. It is my holy city and I am its disciple.

Some time ago we started out talking about radio. We all thought we knew what we wanted. Something of a grown-up, high-speed, digital pair of Dixie cups with an invisible string of radio waves linking them together, right? But light waves are radio waves, and the string can be a beam of light. The light doesn't have to travel freely in space. We can guide it through a fiber just as television comes to many homes in cables instead of through air waves. (Though not mine anymore. I decided to go for nostalgia and bought an antenna.)

Sunlight is a jumble of radio waves. People use sunlight to communicate, with flags, with smoke, with a mirror to flash to rescue aircraft. My wife got to saying, "It's a world of carpeting." But it's really a world of communications. Car horns, clothes, cash register beeps, conversations. Getting bits down from a satellite is pretty constraining, just as is carpeting a house whose inhabitants include people under the age of 8. Wires, cables, and fibers are definitely out. Forget sound—no air to carry it. But there are still lots of possibilities, like conventional radio, spread spectrum radio, and lasers, all of which can operate over an enormous range of frequency, power, and bandwidth. So far we have confined ourselves to frequencies from 0.1 to 10 GHz, which adds up to about 1% of the spectrum of emissions that we know how to transmit and receive.

Satellites, particularly small, LEO satellites, are a key element of the communications infrastructure that will radically alter the way we live and the resources of our earth that we will consume. Our ability to overcome the communications system obstacles we have lived with until now and to produce more throughput for more users with a broader range of applications will determine how big a role our small spacecraft technologies will play in this revolution.

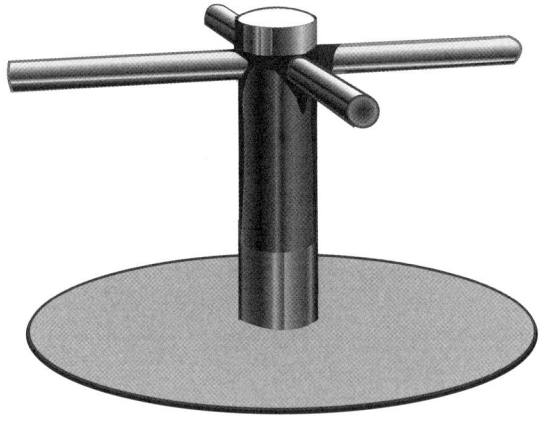

Satellite Antenna.

Chapter 10
Thermal Dynamics:
Tough Talk About Temperature

A short, virtually painless, and occasionally philosophical look at spacecraft thermostatics and thermodynamics

EVERYBODY IN COLLEGE wanted to be a thermodynamicist. Don't ask me why. It's probably because of the big bucks you pull down once you get out, and of course it's very sexy, which is important in school. Naturally, society holds thermodynamicists in very high regard. During WW II, thermodynamicists got special rations of gasoline and tires for their cars so that their vital work wouldn't be seriously impaired by the war effort. And the prestige is a factor you can't measure, but it's certainly there. I have those TD (thermodynamicist) license plates, which gets me a lot better treatment with the Spago Valet Parking. Need a reservation at Four Seasons on a Saturday at 8:30 pm for two right by the window? Just tell the Maitre de Hôte your special guest is a Thermodynamicist. Enough said.

The competition in Thermo classes made my college life Hell. Freshperson classes were chock full, and everybody elbowed for position. Grade grubbing was the rule. Before the six- and seven-year Thermodynamicist programs popped up to get you right into the pipeline out of high school, lots of students spent the whole summer before senior year cramming for the GRE to make sure they got into a Thermo grad school. Thermodynamicist. It's a ticket to country clubs, Lexus dealers, and fancy resorts. Unfortunately, that's what motivated most of the competition.

But I made it, and I can say it was all worth it, even those first two sleepless years of grad school: the 60-hour shifts in front of a micro Kelvin thermostated oil bath and a rack full of HP counters; the internship at an urban combustion tunnel with no budget and a single slit interferometer. Granted, my wife thinks I'm a nerd. Instead of the country club, I swim at the community center with the old folks' low-impact aerobics classes and the toddlers getting their heads in the water for the first time. No Lexus, either. I'm driving a VW, and I don't have time or money for the fancy resorts. I suppose I should've gone into private practice? All my college buds are shopping for wallpaper in Djakarta and throwing Bar Mitzvah parties in Jerusalem. Or they're spending Christmas at the Zugspitze. In bad years, at Aspen. But I do get to design satellites to run at the right temperature. Hey, it doesn't get a whole lot better than that. Who cares about that chalet on Lago Maggiore, a cellar

Optimum locale for studying thermodynamics.

Chapter 10—Thermal Dynamics

full of vintage reds, and sending the kids to Brown and Stanford anyway? The real question is: what is the right temperature for a satellite, and how do you get it there?

No single temperature, of course, is right for everything on a satellite. At one extreme, some instruments might want to be as cold as possible, infrared detectors and very sensitive radio frequency (RF) preamplifiers, for instance. Most batteries want to be in the range 0°C (32°F) to 5°C (41°F). Most electronics want to be around that temperature or a little warmer (10°C, 50°F). Some things want to be warm, like ovenized oscillators and monopropellant thruster assemblies, which are small rockets used for attitude control and orbit adjustment.

Maybe you watched 2001 at an impressionable age, and you think of space as a very cold place. Maybe you think the space sun is very harsh because it isn't filtered by earth's atmosphere, hence the thick helmets on the Apollo astronauts. You are right on both counts, but only sort of.

Assume that a body is isolated. It doesn't gain or lose mass and has no internal source of heat. There are only three ways to change its temperature: conduction of heat, convection, and radiation. If it is "insulated" from these three effects, i.e., no heat can be transferred to or from it, we call its container "adiabatic." Conduction is how you burn the roof of your mouth on cheese pizza (I don't know about you but I hate it when I do that!). The hot cheese gets pressed against the roof of your mouth, conducting heat to your skin, overheating the cells near the surface, damaging, often killing them. That's what we mean by "burning" your finger or your tongue—overheating the surface cells. Our bodies also conduct heat to and from the surrounding air, which is why when it's hot outside, you get hot. Metal is a much better conductor than air, so it cools you off faster, and heats you up faster, than air. This is why a hunk of aluminum at room temperature feels cool to the touch. Room temperature is below skin temperature. Your body is gently having heat conducted away from it by the ambient air, but metal conducts the heat away much faster, so that we perceive it as cool in temperature.

Micro Space Craft

Convection is sort of dynamic conduction. A windy 0°C (32°F) day is a lot colder than a still one. This is because as you conduct, or lose heat, to the air, you warm the air surrounding you. Without convection the warm air near your body actually helps keep you warm. Jackets work because they hold the air around you that you've warmed, keeping you warm. If the air is moving, you constantly lose your warm boundary layer and have to heat up a new one, which subsequently is convected away from you. Isn't convection interesting? No? No problem, since it's pretty meaningless to the heat balance of a satellite anyway, because (brilliant revelation ahead), satellites are not in contact with a fluid medium (i.e., air) to conduct or convect heat to.

A satellite is in physical contact with nothing. Think about that. Nothing touches it. An island in the truest sense. Simon and Garfunkel made millions off this concept! This solitude leaves only one heat transfer mechanism available to a satellite: radiation. The air is cold, I'm told, at St. Moritz in February, but the Alpine sun is warm. Its radiation warms you. Toasters radiate heat to bagels, and a whole industry has grown up around that phenomenon. One very civilized aspect of the "Tube," the London Underground, is the radiant heaters above the platforms, which warm you even though the air in the tube in winter time is cold and damp.

On a clear night, the earth's surface looks up to black space and loses heat by radiation, which makes freezing to death in Sacramento in May a real possibility, even if daytime temperatures are over 30°C (86°F). Desert areas are particularly cold at night because no cloud cover reflects radiated

Convection in action, Chicago-style.

Chapter 10—Thermal Dynamics

heat back to earth. The air is dry and "light," meaning no water vapor helps prevent radiation of heat to space.

With no air to conduct to, and nothing to blanket it from the nearly absolute cold of black space, a satellite lives in the ultimate radiatively dominated condition. How much heat can you or your satellite lose radiating to black space? Where is HAL when we need him!

A thermodynamicist named Stephan noticed that the rate of heat loss from a hot object to its cold surroundings by means of radiation is proportional to the difference of their respective temperatures, each taken to the fourth power, or

$$Q = s\, e\, A\, (T_1^4 - T_2^4)$$

where Q is the heat flux, s, e and A are constants, and T_1 and T_2 are the temperatures of the body and its surroundings measured in absolute degrees, meaning degrees above absolute zero.

In the Fahrenheit system, room temperature is about 560 degrees above absolute zero. (Absolute degrees in this system are known as degrees Rankin.) In the Celsius system, where absolute temperature is referred to as Kelvin (*Note:* Not degrees Kelvin. This strange quirk of calling the temperature units "degrees" finally got axed in the Systeme International (SI) units system.) 295 Kelvin is comfy though not cozy.

Enjoying radiant heat.

The constant "s" is known as the Stephan-Boltzmann constant. One of those fundamental constants like the speed of light in a vacuum, the charge on the electron, or the amount of time it takes after a traffic light turns green for the teen racer in the black, winged, magged, and sound-systemed 5 liter Mustang behind you to honk his horn. "A" is the area of the object that is radiating, and "e" is the emissivity, a measure of the radiation efficiency. Objects with efficiency near one are efficient radiators. Near zero they are poor radiators, which makes them good radiation insulators.

To give you some feel for the numbers, a one-meter square (about 10 square feet) object at room temperature (25°C or about 78°F) that's radiating to black space loses heat at a rate of about 450 watts. That's about the heat flux of a clothes dryer in the low setting, one side of a two slot toaster, or what the sun radiates onto you if you are in LA, fashionably dressed in nihilistic black clothes, and walk over to your therapist's office at noon.

In a gross sense, satellite temperature control amounts to control of the radiation flux. Of course, in a gross sense, all that's required to make an atom bomb is to bring together a supercritical mass of fissionable material. In both cases, and in a surprisingly large set of similar technical undertakings, a lot of work has gone into actually doing that which we so glibly conceptualize.

Two not so small complications to figuring out a satellite's temperature are commonly referred to as the Sun and the Earth. If you expended a significant fraction of your life wondering if the coincidence that the solar flux (heat flow per square foot or square meter) onto our bodies on a sunny day is about equal to the radiative loss from our bodies back to space on a cloudless night amounts to a proof of the existence of God, let me spare you further idle speculation. This rough equality is true, but not surprising. Our bodies cannot need to keep themselves very far in temperature from the temperature of our earth. Otherwise, we'd be incapable of survival and die of hypo or hyperthermia.

Note the significant corollary that our bodies cannot want to be at exactly the same temperature as our surroundings or the heating,

Chapter 10—Thermal Dynamics

ventilating, and air conditioning (HVAC) industry would collapse. Thus, while I may destroy an argument for a great cosmic plan, the same argument asserts that a direct link exists between Creation and HVAC, which may be theologically unique.

The earth's temperature is itself driven by the heat it receives from the sun and the heat it loses to space. Thus, the earth's temperature is that at which the daytime heat input is about equal to the night time heat loss, our bodies' temperature is close to that of mother earth. Voilà! Your radiative heat losses and gains, like the earth's, should be pretty close to each other. An interesting aspect of both our own comfort and that of our satellites is that small temperature shifts affect heat loss significantly, because the heat loss is proportional to the temperature taken to the fourth power. A 5°C shift, only about 1.6% of the absolute temperature, results in a 7% shift in the heat flux. Thus, if for some reason a satellite gets 5°C warmer, it loses 1.07 times more heat at night. But it also gains a little less heat during the day, both of which tend to cool it off.

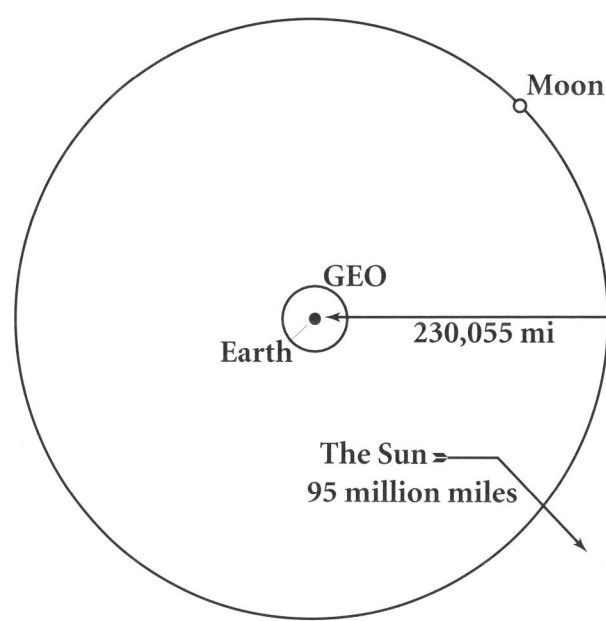

GEO orbit, moon, and the sun about 50 feet off the page, to scale.

Micro Space Craft

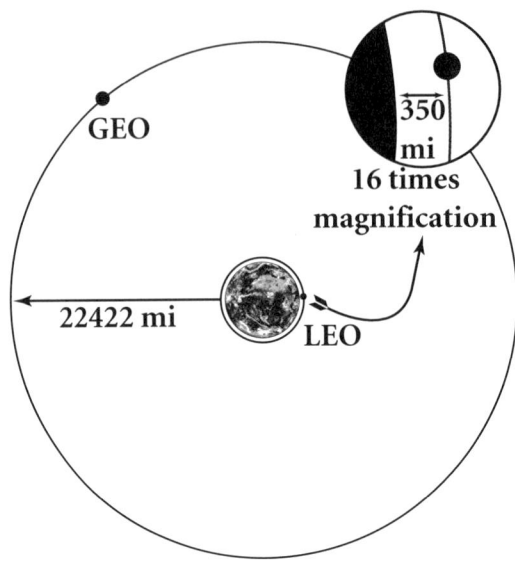

Earth, GEO, and LEO to scale.

The earth's effect on a satellite's temperature is a bit more subtle. At very high orbits, like geosynchronous orbits, the earth forms a small part of what the satellite sees, or radiates to. Its temperature is determined mostly by the flux from the sun, about 1,300 watts for each square meter, and its radiative flux to space, about 450 watts for each square meter.

Just like the earth, the satellite absorbs sunlight only over the part of its surface exposed to the sun, but it radiates to space from its entire surface. This ratio is (assuming a spherical satellite, which for some reason is a really funny punch line in a lot of obscure thermodynamic jokes) exactly 4. Thus the disparity between heat loss and heat gain is actually about 1300 W in compared to about 1800 W out. The satellite will be cold. The bright metallic, usually gold-colored coatings so commonly seen on geosynchronous satellites are thermal trim materials. These metallic surfaces are very efficient absorbers of solar radiation, but radiate weakly at the temperature of the satellite body. Thus they tend to warm the satellite. The exact amount and type of this surface coating, its precise placement, and its degradation in the space environment is a significant element of any geosynchronous satellite's design.

Chapter 10—Thermal Dynamics

At low earth orbit (LEO) the job is a little easier, because the earth is close by, relatively speaking. Close this book and hold it at your maximum arm's length. Really reach way out there. It is about as large from your eyeball's point of view as the earth is from a geo-synchronous satellite's point of view. Hold the book as if you were reading it normally (even though you're not at this point). Then bring it very close to your eyes so that you actually touch it to the tip of your nose, as if you were extremely near sighted. It is about as large to your eye as the earth is from a LEO satellite's viewpoint. Close! So what, right? Because the earth occupies so much of its view, the satellite is strongly radiatively coupled to the earth. Remember the area term in the radiation equation. About half of the satellite's surface area sees the earth. If the satellite tries to get warmer than the earth, it radiates heat to the earth and cools. If it tries to significantly cool itself, the radiation flux from the earth warms it.

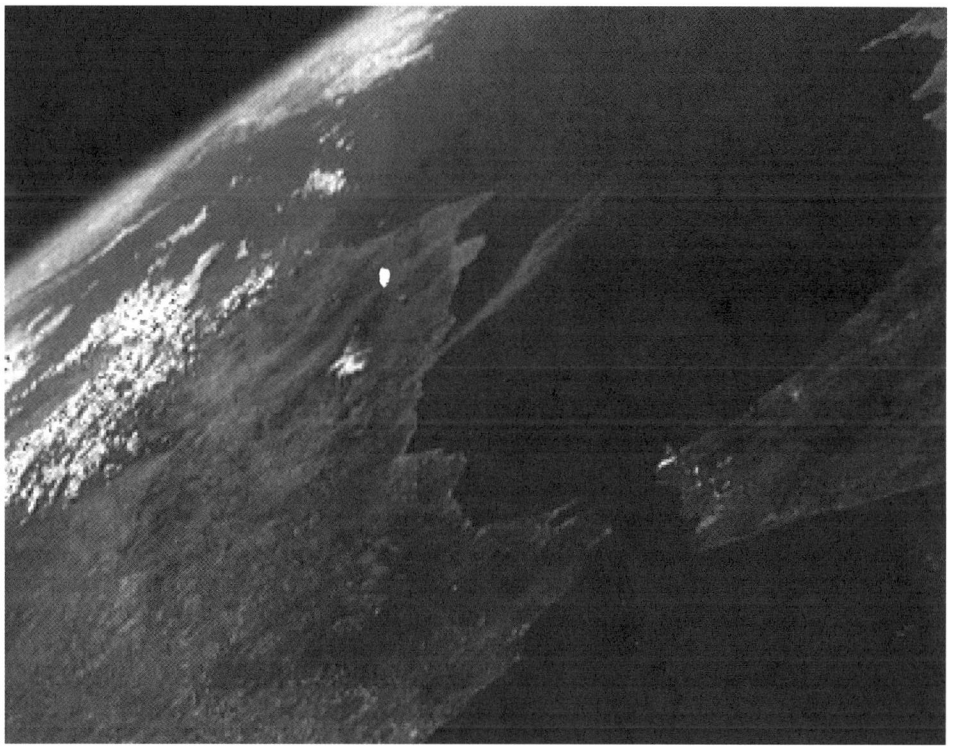

The earth, seen from LEO, dominates your field of vision.

117

Micro Space Craft

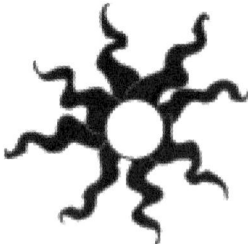

Just as it is no major coincidence that our bodies are near in temperature to that of our surroundings, which in turn are in equilibrium with the radiation in from the sun and the radiation loss to space, LEO satellites are hard pressed to reach a significantly different temperature than the earth. On average, the earth is a little above freezing, about 10°C or 50°F. And that, not so surprisingly, is the temperature of most LEO satellites.

"For that my genius son went to school for 22 years?" I can hear my father saying. To find out that no matter what you do you end up at 10°C? Hey, some kids turn out, some don't. My family doctor went to school for 23 years and every one of his patients eventually dies. But a lot of interesting stuff happens in the span between birth and death. Similarly, the variations around those 10°C averages are pretty important.

For example, what if we have a component we'd like to maintain at lower temperature than the rest of the satellite? What you do is first point the satellite at the sun, so that the object can fly in the satellite's shadow. Then thermally isolate, i.e., insulate, the object so that it can't easily conduct heat to or from the rest of the satellite. Create some radiation baffles, or barriers, to stop the warm satellite from radiatively coupling to the hopefully cold object.

Sun-pointing satellite with shielding.

These barriers can be just curtains or walls coated with a low e (emmissivity) material. While you're at it, don't forget to similarly isolate the cold object from "seeing" the earth. Finally, given that it can't see earth or sun, expose it to what's left, namely black space. Depending on the care with which the insulation and radiation barriers are constructed, objects can be cooled to as much as 100°C (180°F) below the earth equilibrium temperature.

Another small detail that provides job security for us thermodynamicists is that while a 20°C (36°F) fluctuation is small in absolute

Chapter 10—Thermal Dynamics

temperature terms—less than 10%—it makes a big difference to the satellite's components. For instance, Nicad batteries are very happy at 10°C (50°F), but they are a bit cranky at 30°C (86°F). It could be dangerous to charge them rapidly at such a warm temperature. Plus, satellites move a lot of heat around. For instance, turning on a transmitter causes local heating around the transmitter as it dissipates power. So a lot of work can go into analyzing the precise temperature different parts of a satellite will reach.

Small, low cost satellite programs avoid some of these complications. The most basic approach is to thermally short all parts of the satellite together. Not letting any one part be insulated from the others prevents it from getting too hot or too cold. Since it is easy to transfer heat over short distances, thermally shorting the pieces of a small satellite amounts to just bolting them together, assuming everything is made of a good heat conductor like aluminum. The 50 kg ASAP payloads, like the AMSAT cubes (shown below) use this approach. They mostly do not get colder than 0°C (32°F) nor warmer than about 15°C (60°F).

Which is not to say that the solution is as trivial as a few bolts. Small satellites have been flown with thermal blanketing, usually to reduce radiative heat loss on shaded areas of the satellite. They have louvers to compensate for variations in solar heating during interplanetary trajectories away from or toward the sun, with heaters to ensure that critical components are never too cold. Current designs even exploit heat pipes to aid in transferring heat away from highly stressed components.

Some satellites must turn their broad sides away from the sun during highly illuminated orbital seasons to reduce spacecraft heating, and local heating plays a key role in some satellite components. Gravity gradient booms can create oscillations in the satellite pointing attitude if their illuminated surfaces warm and stretch while their shaded surfaces cool and shrink. At the component level, without air to conduct heat away from microprocessors and other high performance components, they can get very hot. Many flight circuit boards have a special copper layer inside them to sink heat away from these components.

Micro Space Craft

AM SAT cube.

Interesting, isn't it? A satellite is its own microcosm, isolated from the universe except for the radiative exchange of energy. When you think about it, the solar collectors are radiation receiver/absorbers, the radios are highly tuned radiators, and the satellite absorbs and emits radiant heat. Even remote sensing instruments like telescopes or sensitive radio receivers are just other means to absorb incident radiant energy.

On earth we take our rich environment for granted. Heat is conducted to our surroundings. When something gets real hot or real

Chapter 10—Thermal Dynamics

cold, radiation and convection kick in to help smooth things out. Air brings us oxygen and aroma, a sensation of speed and some momentum, at least to blow our hair around. It's a medium for the pressure fluctuations that transmit sound to our ears. Air provides us a stable thermal environment to exchange heat with. The satellite's environment is much more spartan. With no molecular transfer, it lives submerged in a gossamer sea of photons. Radiation is the only contact the satellite has with the outside world. Thermal design regulates photon absorption and emission, and the distribution of the heat gained from those photons through the satellite's body.

Chapter 11
You Got an Attitude, Buddy?
A Primer on Small Satellite Stability and Control

IT'S A FACT OF LIFE. Most things have an attitude. That's why I retired my beloved cat, Providence, to a horse farm in the Midwest. Small satellites (you knew I'd work around to them) have attitudes, too. Attitude determination and control are key features of satellite design and they're a special problem for small satellites. Antennas like to point at the ground for uplink and downlink. Solar panels like to point at the sun. An astronomical telescope may want to point at a star cluster or along a set of celestial coordinates. For sky survey astronomy missions, detectors like to spin slowly and scan the whole sky. A camera or directional antenna may want to point at Lima, Ohio, if that's where the news is.

Believe it or not, on the ground, in the Washington winter fog on New Year's Eve, at 2:00 a.m. after sampling every beverage at three different parties, your attitude control still has a lot more going for it than most small satellites'. For instance, you know where down is (the direction your house keys just went into the snow). For a satellite, down is a difficult concept. If a satellite were to drop its keys, where would they go? Granted, the topic is academic. We don't fly satellites with keys, but since when has that stopped anyone from studying an interesting problem? Satellites, having hypothetically dropped their keys, should have absolutely no problem. First, there is no snow in orbit to lose the damned things in. Second, the keys continue to move along with the satellite and, from the satellite's reference frame, go nowhere. But the satellite senses itself to be in a zero-g, no gravity, environment. So down is tough.

Micro Space Craft

Can you walk a straight line at 2:00 a.m. New Year's morning? How about almost straight? Again, you are better off than satellites. Telling which direction you are going in a satellite is subtle. Very subtle. No wind. No sequential ads for chewing tobacco or Burma Shave, and no roadside to post them on. On the other hand, no Ladybird Johnson to get all upset about roadside signs. Lots of plusses in space. Lots. But forward is tough. From our satellite's point of view, it's not moving at all. Next time you fly on an airplane, walk into the restroom and close the door. How can you know how to point in the direction the airplane is going? Think of it as another step in the slow process of gaining empathy for satellite guidance and control.

For these and other reasons (like the fact that in 1957 getting into space, period, was more important than knowing where up, down, or frontwards was), the first satellites were unstabilized. Like Exhibit A here:

Un-stabilized

Sputnik—Khruschev's first attempt at Star Wars.

What Sputnik and our early satellites did about the attitude problem was—I can relate to this—they ignored it. They put a bunch of antennas in all directions, which made it easy to build neat models of satellites. Get a golf ball, some Elmer's® glue, and some colored golf tees. Glue the tees to the ball all over its surface, and, voilà, Sputnik. The satellite is clueless about where it is pointing and doesn't care. It worked fine then and it works fine now. The simplest solution is to ignore the problem. If you can do that, you can save money, time (which is money anyway, isn't it?), and payload weight for important things like ant farms and commemorative stamps.

Read dress-for-success books. They'll tell you pointing down is a very authoritative gesture. Looking down is handy for things like tying shoes and finding those damned keys. And pointing directional antennas down at the

Chapter 11—Stability and Control

earth is handy for receiving weak signals from downed airplanes blown off-course en route to Red Square, or taking pictures of clouds over the earth to show on the six o'clock news. There is only one way to be unstabilized—like there's only one way to be dead. But there are lots of ways to be alive, for example, George Bush, George Harrison, Boy George, and there are lots of ways to point down.

Gravity Gradient (GG) uses the fact that the earth's gravity gets weaker as your distance from the earth increases. Thus, a satellite with a weight stuck out on the end of a long boom sort of hangs on the end of the boom, like little iron filings all lining up in a magnetic field. The satellite doesn't care if it is down and the weight is up, or if the weight is down and the satellite is up, so a method to flip it over is needed. But it's vertical. The variation in gravity over the length of a boom, 20 to 100 feet, is minute, and so is the gravity gradient torque. About the equivalent of the torque of a lightweight (flea weight) flea standing on the seat of a playground seesaw. Because no other stronger forces are present to upset the system, this minuscule torque orients the satellite vertically.

Gravity Gradient

Hanging out on the long boom of the gravity gradient (GG).

Gravity gradient satellites are popular because the stabilization is passive. No propulsion or gyroscopic systems are needed to maintain the vertical orientation. But the stabilization is weak and hence small oscillations about the vertical, up to +/-10 (degrees), are common. If anything starts the satellite rocking, like its initial separation from the launch vehicle, GG has no damping to attenuate the oscillatory seesaw motion. Thus dampers need to be added. We'll get to dampers later. Suffice it to say they add an incremental complexity to an otherwise simple stabilization solution.

I could go on. And on. But let's wrap up the first part of this primer with one more passive stabilization technique. Next time we'll take a look at systems that require the satellite to actively do things like spin itself or spin wheels or turn magnetic fields on and off to use the earth's magnetic field to create torques.

There isn't no wind in orbit (run that through your grammar checker). Just very little. Air density decreases exponentially with altitude. At space shuttle orbit altitude of 300 km, or 160 nautical miles, aerodynamic drag is about 100 times stronger than gravity gradient effects. By moving the aerodynamic center of pressure behind the mass center of gravity (in other words, by adding the equivalent of tail feathers to an arrow) the satellite points into the wind. Since the wind is created only by the motion of the satellite, like creating wind in a car by driving forward, pointing into the wind is the same as pointing in the direction you are going. In warm weather, dogs riding in cars often use the same technique to point their noses out the window in the direction they are going. The relationship between these applications of weather cocking is determined only by moving to the East (Tibet, not Baltimore) and thinking hard for 20 years. While you dial up your travel agent, the figure below shows an aerodynamic stabilized satellite.

Aerodynamic

Aerodynamically stabilized satellite.

The Chemical Release Satellites built for the Air Force are examples of this design approach. Stabilization pointing frontwards is needed to find the satellite from ground or airplane-based telescopes. The tail fin doubles as a radar reflector to increase the radar cross section of the satellite to make initial acquisition easier. An optical beacon (strobe) is carried that the stabilization system keeps pointing frontward. Don't you feel Domino's Pizza and Federal Express have cheapened this neat concept by using it for their delivery vehicles? Write your congressperson.

The next part of this subject, spin stabilization, requires deep thought, possibly 30 years in Tibet. In the meantime, if you see Providence, tell her all is forgiven.

Chapter 11—Stability and Control

Active Control

Besides expressing tremendous empathy for the difficult conditions satellites live with every day of their lives. Chapter 10 reviewed many, though not all, of the popular stabilization systems that require no active control of

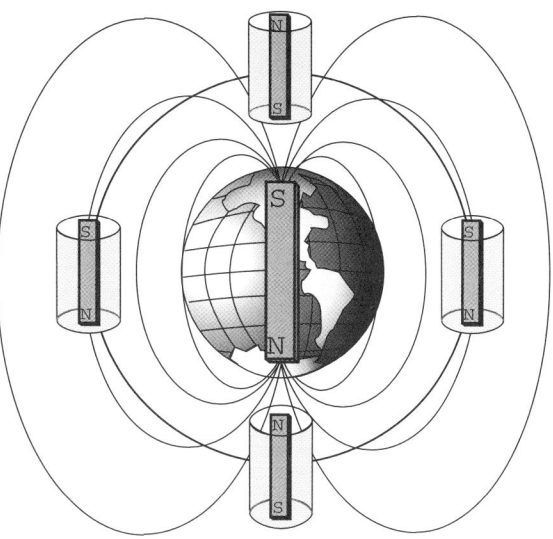

Magnetic stabilization.

the satellite. These included no stabilization (heavily favored by Existentialists), gravity gradient, and aerodynamic methods. In addition to these, a number of schemes using permanent magnets have been tested.

Permanent Magnet

Permanent magnetic axis alignment of satellite.

Using a permanent bar magnet fixed in the satellite, one satellite axis is forced to align with the earth's magnetic field lines. The sketches above and below illustrate this stabilization, which causes a twice-per-orbit flip in polar orbit.

Permanent magnet stabilized satellites have no stability around the field line and tend to roll about it, which precludes pointing at the earth. One exception has been exploited by Amsat. A dipole antenna can be aligned with the magnet axis and hence is aligned with the magnetic field lines. Except for near the magnetic poles, these lines are roughly parallel with the earth's surface so that the side lobes of the dipole are earth-oriented. This provides about 2 dB of antenna gain which, for the price of a bar magnet, could be a bargain.

Micro Space Craft

I hesitate to leave passive stabilization. It works; it's inexpensive; but it tends to be overlooked and underrated because big satellites have largely stopped using it. Also, it helps rid your organization of pesky digital and automatic control engineers who are, at best, expensive, and at worst a negative influence on corporate decorum, what with their drinking, personizing (we used to say "womanizing"), and so on.

But I digress. If you find the beauty of active control beckoning you from beyond the ken (where it's always been for me), then read on.

The most complex active control configuration to implement is, paradoxically, the easiest to relate to. Shown below is a three-axis stabilized satellite.

If you look at it from, say, a chaise lounge fixed in space, preferably close to a blue-green ocean inlet dotted by wind surfers plying a light trade wind from the East, a three-axis stabilized satellite would maintain a fixed orientation relative to you. If your chaise lounge were near an x-ray pulsar (rotten luck, I'd say) or an interesting interstellar gas cloud, you could appreciate an important feature of three-axis stabilization: it can stare at a single point in space.

Note that, without some moving around, it isn't going to maintain a fixed orientation relative to the earth. Imagine the satellite right side up over the North Pole. Its legs, if it had any, would be pointing down toward the earth. It is fixed in

The three axis stabilized satellite, right side up at the North pole, would appear upside down (head on the ground, legs in the air) at the South pole.

Chapter 11—Stability and Control

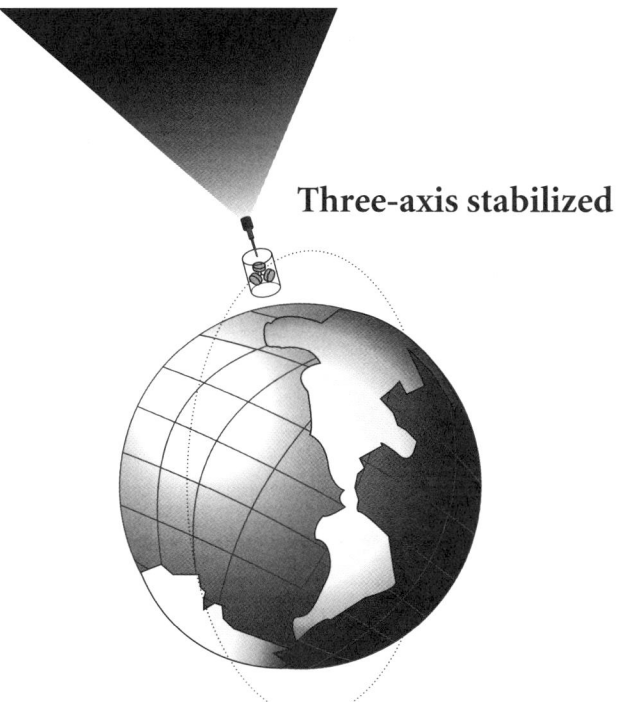

The three-axis stabilized satellite, right side up at the North Pole, would appear upside down (head on the ground, legs in the air) at the South Pole.

space, so as its orbit brings it around the South Pole its legs are now pointing away from earth. From an earth observer's point of view it is now upside down.

On the other hand, if you are a satellite in orbit around the equator, your up-downness relative to earth is unchanging, which is the preferred way to stabilize the antennas of modern communications satellites. In continuous equatorial orbit, their three-axis stabilization allows them to point continuously at their service area.

Small disturbances, however, tend to rotate the satellite away from its desired attitude. This small motion is usually sensed by gyroscopes. The most common way to deal with the rotation is to design a three-axis stabilization system that uses reaction wheels. A small rotation forward is countered by spinning an internal wheel in the same direction as the sensed upset rotation. The equal and opposite reaction of the rest of the body (conservation of momentum) causes the satellite to slow its forward rotation proportionately to the amount the wheel spin speed is increased.

By continuously sensing rotations on all three major axes and controlling the spin rates of three wheels, the body can be held inertially fixed. For a while.

129

Micro Space Craft

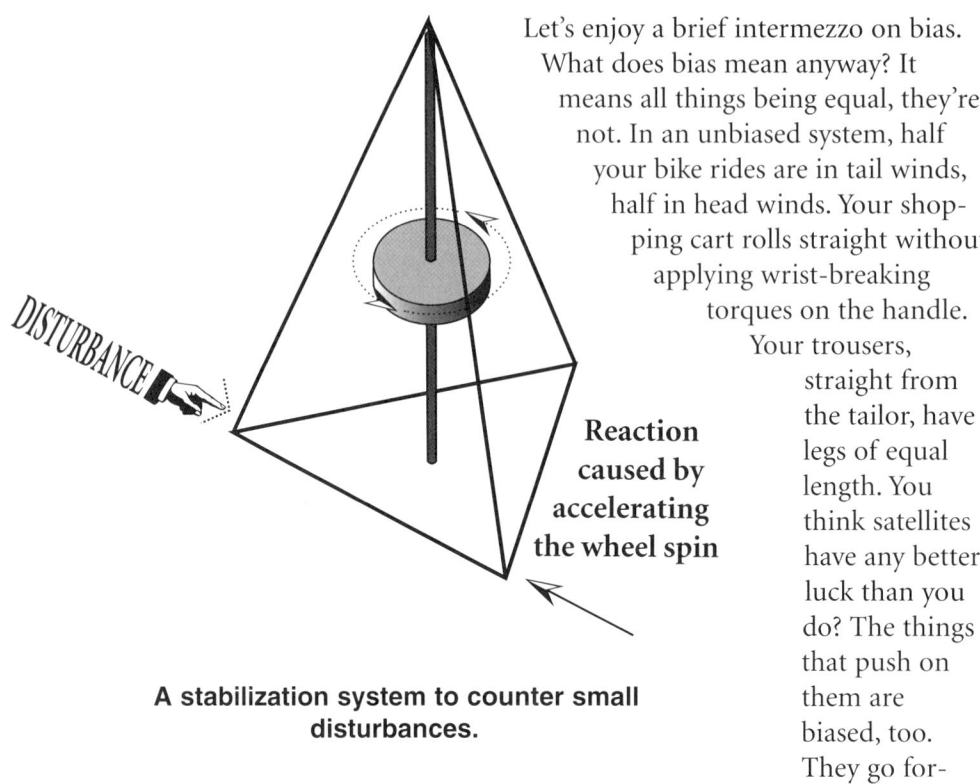

A stabilization system to counter small disturbances.

Let's enjoy a brief intermezzo on bias. What does bias mean anyway? It means all things being equal, they're not. In an unbiased system, half your bike rides are in tail winds, half in head winds. Your shopping cart rolls straight without applying wrist-breaking torques on the handle. Your trousers, straight from the tailor, have legs of equal length. You think satellites have any better luck than you do? The things that push on them are biased, too. They go forward more than back. Or back more than forward. The sign (+ or -) doesn't matter. What matters is that the disturbance torques applied to satellites are biased. Why? Sometimes because satellites aren't symmetrical. Sometimes because life is like that.

Living with the hand Murphy dealt us, in this case, requires handling bias. If external forces keep pushing the satellite forward, we have to keep speeding up the forward spin on the reaction wheel to counter them, faster and faster. The wheel is spinning forward and if we slow it down, equal and opposite forces push the satellite forward. So we can't slow it down. Every time nature gives the satellite another push forward, we have to increase the forward spin on the wheel to counter it. But this can't go on forever, because the reaction wheel can spin only so fast. When it's spinning that fast, it is called saturated. Now what? The wheel is spinning forward as fast as it can and we get yet one more disturbance that pushes the satellite forward.

Chapter 11—Stability and Control

Kind of leads you right to the brink, doesn't it? Our satellite is caught between duty (pointing in one direction) and the immutable forces of nature, biased disturbance torques. You might wish to stop here so as not to ruin the ending when this is made into a 49-week docudrama on PBS and everyone in the office talks about it every Monday morning.

Or you might want to hear about propulsion. A lot of three-axis stabilized satellites have lots of very small rocket motors, around 1 pound force of thrust (for you metric sticklers, about 4. 5 Newtons) distributed around them. Probably a hundred times you asked yourself why giant satellites have these tiny motors. Okay, maybe once or twice you've wondered. Even if you've never wondered, I'd say it's an interesting question and probably worth trying at cocktail parties and office lunches.

The boss decides the group needs a little more adhesion so all 37 of you pile into 19 Ford Escorts, minivans, motorcycles, and Porsches and head out to the local Taco Shop, which used to be a health club with a one-lane lap swimming pool. You are seated at a table three feet wide and 25 meters long and your group bonding experience amounts to facing one person you know only too well and have been trying to avoid for six months. The guy you've been wanting to BS with about that PBS docudrama is down where the diving board used to be. You're in the four feet. Well, here's the answer. Ask that true zero whom Fate has cast into the middle of your bonding lunch what all those little bitty rocket nozzles are doing on big, three-axis stabilized satellites. Here's the answer:

When that forward spinning momentum wheel saturates, or gets near to saturating, or just plain gets tired of spinning pretty fast day in and day out, one of these rocket motors is fired, which nudges the satellite the opposite direction from the biased disturbance forces. In our case, where natural forces are pushing the satellite forward, the motor is used to push the satellite backwards. Now the wheel has the opposite problem—the satellite is rotating backwards. The wheel slows its forward rotation to counter this backward motion. By applying small impulses with the motor and seeing the reaction wheel respond, the wheel can be returned to about zero spin rate. And that is one of the things small rocket motors on big satellites are for.

Another way to desaturate momentum wheels that doesn't require propulsion is to use magnets. A satellite in earth orbit is influenced by the earth's magnetic field and can generate torque using this magnetic field. An electromagnet coil is turned on when its orientation relative to the earth's field results in a torque in the right orientation to rotate the satellite in the desired direction to desaturate a reaction wheel.

The magnetic system is not always the solution to the desaturation for several reasons. In the very high orbits used by modern communications satellites, the magnetic field is weak and highly variable, making it very hard to use. Even in low-earth orbits where the magnetic field is stronger, the field lines are not always aligned well for producing the torques required to desaturate the reaction wheels. Despite the ability to use the magnetic field for the stabilization task, rocket motors could be necessary for trimming the satellite's orbit. Designers may opt to get double duty from the motors, using them for so-called station keeping propulsion and for stabilization. Both propulsion and magnetic torquing have been successfully employed with reaction wheels to achieve three-axis satellite stabilization.

Spin Stabilization and Conclusions

Now all can benefit from the highly enriched Pearls of Wisdom (POW) I gleaned during endless Popsicle® breaks taken at my previous place of toil with the sharpest engineer who's ever condescended to talk with me, Richard Warner:

POW #1: Thermal vacuum chambers are available for simulating space thermal environments. Anechoic chambers can show you how antennas are going to work on orbit. Shake tables and centrifuges verify structural strength. But there are no 0-gravity chambers to play with torque-free motion.

There it is—the wisdom of the ages right here. Who goes into a thermal vac test and changes nothing? Who does shake and antenna

Chapter 11—Stability and Control

tests and doesn't see anything worth improving? Not me; probably no one. But we routinely design and build complex satellite control systems with various sensors, controllers and actuators without any testing besides computer simulation and analysis. We expect the resulting systems to stabilize, point, and track satellites because there is no room into which we can throw the satellite and have it rotate and translate freely without disturbances from, say, hitting the floor or the walls. Typically, we can simulate sensor environments and measure actuator response. But no one can watch a gravity gradient stabilized satellite approach equilibrium in a laboratory. This unfortunate fact is, by the way, quite often used as a lever by those scheming G&C engineers to extract ever more expensive computing equipment and salaries from their nearly bankrupt employers. Which brings me to **Pearl of Wisdom (POW) #2:**

Every silver lining has a cloud.

Armed with our newly achieved enlightenment, spin stabilization is easier to treat. Torque-free motion is the way an object behaves without external disturbances like friction from touching surfaces and air resistance. In torque-free motion, a spinning body spins forever. Its spin axis—the direction the axle of a spinning wheel points— would never change direction. Of course, even if an object doesn't spin, in torque-free motion and without initial rotation rates, its axes never change their orientation in inertial space.

The difference between the torque-free motion of a spinning body (for example, a spin-stabilized satellite) and a non-spinning body (an object in orbit without any rotation) is one of those subtle physical things that many people would rather quit physics in high school than really get to know and love. Instead, they become bond traders and surgeons and CEOs of major corporations. But from their private jets and expensive homes overlooking Puget Sound, they know that something, some subtle physical insight, is missing from their lives. It is this void, Freud tells us, that motivates Donald Trump to put his name on skyscrapers, used 727s, and blimps. You can avoid all the time and trouble involved in amassing these

Micro Space Craft

uncountable millions, only to see a substantial fraction of them squandered on taxes and Jaguar XJ-Ss, on diamonds, and single-malt Scotch.

It is much more elegant simply to understand that when a torque impulse is applied to a non-spinning body, the body begins to rotate in whatever direction the torque has twisted it. Thus, a satellite initially pointing at, for example, the sun, soon loses this orientation after a torque is applied. It slowly rotates with no particular orientation, which, if you really want to look at the sun, is counterproductive at minimum. The situation is illustrated stunningly below.

It is even more elegant to realize that a spinning body, suffering a small torque impulse, responds only by having its spin axis, the direction its "axle" is pointing, change by a small, fixed angle. Thus, if a satellite is initially pointing at the sun and a disturbing torque impulse is applied, it still points in a fixed direction, deflected a small, unchanging amount away from the center of the sun. The more angular momentum (faster spin and higher spin axis inertia) possessed by the body, the smaller this disturbance angle is. The illustration shows the difference.

Now, isn't that easier than spending your time buying and selling major corporations? What good is a corporation once you buy it? What if you bought USX? Nobody really knows what USX makes

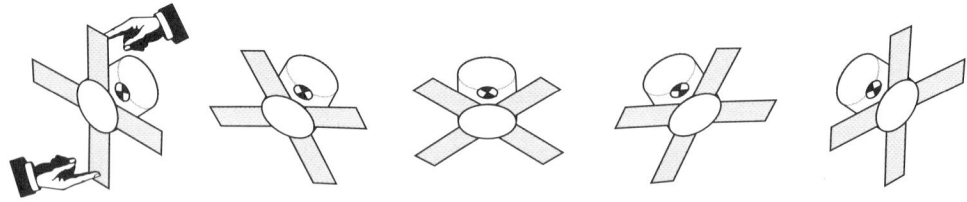

A non-spinning body, when subjected to a torque impulse, will begin to tumble and continue to do so.

Non-spinning satellite tumbling after torque is applied.

Chapter 11—Stability and Control

anyway, and it really is only interesting cocktail party talk for a short while. Then you have to sell and buy again, maybe AMF this time. Then take up bowling. Spinning satellites, on the other hand, are neat and have real value.

This value stems from several attributes. Amsat's OSCAR IV and several following it spin so that their antennae, oriented on the spin axis, present gains to the earth when the satellites are near apogee. Thus they save power and reduce antenna requirements. Also, they all avoid a satellite's worst enemy—tan lines. Your know that if you fall asleep in the sun, your tan is uneven. Horrors! A true tanning buff (no pun there) wants to rotate slowly like a rotisserie so that he or she is evenly exposed to the sun. Satellites like this too. It keeps them from getting too hot on one side and to cold on the other. Thus, spinning reduces temperature gradients while helping hold a constant, inertial attitude.

ALEXIS, a spinner with a different attitude in many ways, always faces the sun. Its sun-facing surface is mostly solar panels that don't mind, in fact rather enjoy, continuous solar exposure. Its payload, on the other hand, hates to face the sun. A spinning satellite can

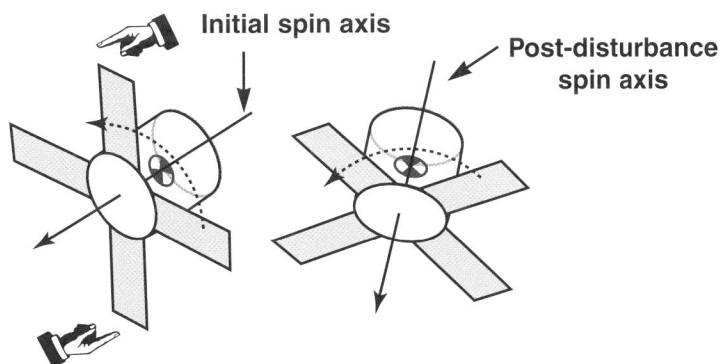

A spinning body, when subjected to a torque impulse, will precess its spin axis to a new orientation and maintain that orientation.

Spinning satellite with small deflection after torque impulse is applied.

provide this yin/yang experience because if a disturbance torque steers it a few tenths of a degree away from the sun it can be slowly dragged back by applying torques from its magnetic torque coils' interaction with the Earth's field. There is no hurry to make these corrections. After suffering a disturbance torque, a non-spinning ALEXIS would slowly, continuously rotate. Eventually the payload would face the sun and become unhappy, confused, and possibly damaged while the solar panels go into shadow and stop making electricity. A very bad state. But since it is spinning, the same disturbance torque now creates only a fixed, small pointing error, say 0.1 degree, and that's all. If the G&C system (like the engineer who built it) is out to lunch 22.3 hours a day, No Problem! When it awakens, it does not find a disoriented satellite, just a (ho hum) 0.1 degree pointing error. It corrects this error by applying an opposite torque with its magnetic coils.

One of the potential disadvantages of magnetic torquing is that the correct torque is not always available. Because of its stiffness, a spinner can afford to wait for correction opportunities even if they occur only a few times a day.

Spin stabilization is used in a range of applications. Spinning up rocket upper stages before firing makes the rocket point in a nearly constant inertial direction despite inevitable disturbance torques produced by small thrust asymmetries. Geosynchronous communications satellites, particularly from Hughes, have a spinning body that provides a stiff platform for orienting highly directional antennas. The antennas themselves cannot spin because they have to stare at a region on earth, so Hughes uses a large bearing and the antennas sit on a de-spun portion of the satellite.

Another variation on the spin-stabilized satellite is to move the spinning portion into a small box somewhere on the satellite and leave the rest of the body unspun. This configuration, shown below, has the same response to a disturbance as the all-spinning satellite.

The advantage is that only a momentum wheel inside a housing, which looks temptingly like a computer hard disk, spins while the rest of the satellite does not. Configuration power needs to be sup-

Chapter 11—Stability and Control

Momentum Wheel

Stabilizing effect of momentum wheel.

plied to get the wheel going. This power varies roughly with the amount of stiffness you need, which in turn depends on the satellite mass, how tightly the attitude must be controlled, and the size of anticipated disturbance torques. To be reasonably stiff, a 250 lb. satellite at 400 mile orbit requires only a few watts to spin a wheel weighing a few pounds.

Conclusions

In this chapter I have irreverently but accurately glossed over every major type of space vehicle stabilization system used today. All without more than a few smirks and furtive grins from the real G&C types I have to deal with regularly. These systems have included:

- Unstabilized

- Passive Stabilized: gravity gradient, aerodynamic, some spinners, magnetic

- Active Stabilized: spun/despun, some other spinner, three-axis

It should also be intuitively obvious, as they say, to the most casual observer that these themes have lots of variations. On a welcome note, in this book you won't see any more attempts to address these myriad concepts. Having lived through this voyage into the virtually unknowable, the only cloud enclosed in your silver lining is the gap that finishing this opus might leave in your reading. Have you tried the yellow pages?

Chapter 12
Memory Systems for Spacecraft
—or
Memory —
What Is It Good For?

— by Richard Warner and Rick Fleeter

ACCORDING TO SEINFELD, the original title of *War and Peace* was "War — what is it good for". Talk about rhetorical questions — no wonder the publisher dumped it. The song (by the group, War) of the same name notwithstanding, War is good for killing your enemies, plundering, raping and absorbing wealth,maximizing your popularity among your legions, and raw ego satisfaction. It's a great technology driver, and paves the way for cultural imperialism and opening up new markets. If War didn't exist, surely Coca Cola or IBM would have to invent it (hark! a war cry of the `60s when we thought they really did.).

Satellite memory systems are good for something too — but somehow they lack the death/sex/violence cachet of War or Pulp Fiction. While the infoworld has focused on computing power — 8086/80286/80386/80486 and Pentium all have become household words — the manipulative power of a fast microprocessor is worthless without data rapidly available for manipulation. In 10 years, the amount of memory we think we need in our computers has leaped from 32k or 64k (k being roughly 1000 bytes) to today's 6M to 64M (M being roughly 1 million bytes). A factor 1000 increase. The case

Micro Space Craft

can be made (though not in this book) that TV ratings need a good war here and there, and that solid state memory like that found in PCs created the small satellite revolution.

To be small and simple, ergo, cheap to build, and to be cheap to launch, microsatellites tend to be in Low Earth Orbits (LEO) from where the earth appears rather hugely close by (an object 12,000 miles in diameter only 300 miles away looks something like a beachball an inch from your eye).

At that range, you can't see a whole lot of the beachball or the earth, and from the microsatellite's perspective, it can't communicate with what it can't see. Flipping over all the cards, most small satellites only see the ground station that links them to the people that paid for them to frolic carelessly in the heavens about 25 minutes per day, coincidentally about the same amount of time most dogs are active.

So what do they do the rest of the time? Of course with dogs, the options are few — eat, lounge, sleep, rip up furniture. With satellites, the amount they can do during the 23 hours and 35 minutes of solitude afforded them depends expressly on how much memory is on board. With no memory, they are like dogs except they're easier on the furniture. They can soak up solar energy and charge batteries, and they can sleep. Now even a cheap satellite in orbit with a ground station will run you a few million dollars, and having that investment sitting idle 98.3% of the time could be equated to spending a couple of $100M on something that would work 100% of the time — like a large geosynchronous satellite in orbit over the equator at your longitude. To be cost effective, small satellites need to be awake much more than dogs, and employing a large on-board memory, they can do that.

Maybe you want to map the whole earth, looking for spectral signatures of oil near the surface or undersea. Ideally, the satellite would remain awake all the time and take hundreds of digital images of the earth's surface in a few key wavelength bands. What happens to all those images until the next pass when they can be sent down to the ground? They get stored in the spacecraft digital memory.

Chapter 12—Memory Systems

Maybe you want to collect status messages from sensors all over the world. You might be monitoring ocean temperature, or just keeping track of the positions of containers on ocean going vessels and trains, or the status of pumping stations and electrical distribution centers. The satellite collects these data to send down on the next pass over the ground station — and meanwhile stores it in the spacecraft digital memory.

Maybe the satellite images the astronomical objects. Many of these are not bright, requiring integration over very long times. Their images are slowly built up through repetitive addition of photon counts into the spacecraft digital memory.

None of these applications are possible without large, fast, low power and mechanically simple devices for mass storage of digital data. The most capable small satellites can downlink about 128 kBytes (about one million bits) of data per second. With an average of 1500 seconds of contact time per day, 2 Gbits (2 billion bits) of data is typically targeted as a data storage capacity, since that's all the data which could be downlinked in a day's worth of contact with the satellite.

History

Classically, satellites either had no storage or they used electro-mechanical devices like tape recorders. There are still many satellites in orbit, quite costly sophisticated ones, with less memory than an average home computer. Because satellites evolved before large memories were available, their missions have been built around real time control and data relay. They require no more memory than does a telephone or a television set since they do not have the job of storing data, but rather relaying it as quickly as it is received. They do not execute stored commands, but rather respond to real time commanding from the ground.

Small satellites, because of their tight volume, mass, and cost constraints, and because they have found wide application in providing digital data store and forward (mailbox and e-mail) service, have

Micro Space Craft

really driven the development of the newer generation of solid state memories. Working at AeroAstro, I (RW) have been involved in the development of digital memory systems for several small satellites.

What's Available in Satellite Memory Devices?

The classic data storage device for spacecraft for decades has been the flight tape recorder. These were reel-to-reel devices like you would see in a mainframe computer installation, complete with motors, capstan drives, springs, and guide wheels.

Its main disadvantages include large mass (tens of pounds), high cost (close to $1M), reliability problems due to the moving parts,

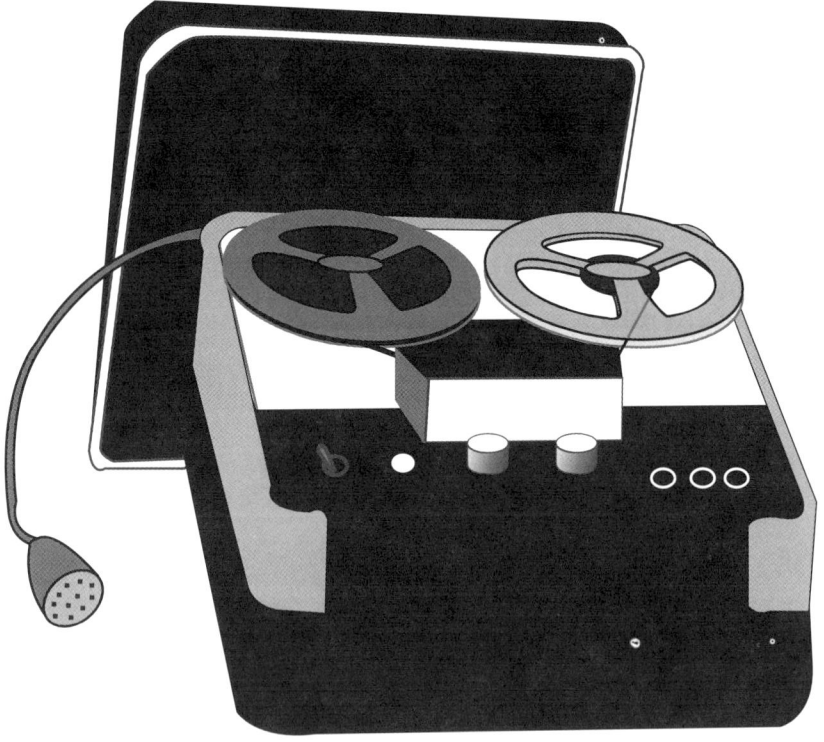

A reel-to-reel storage device from the Jurassic Era.

Chapter 12—Memory Systems

and lack of flexibility in terms of access to the data — random access to specific portions of the recorded data is virtually impossible. Anyone who has tried to find their favorite track in the middle of a cassette tape knows this problem! Other moving media devices have been considered: computer hard disk drives, magneto-optical disk drives, and VCR-type cassette drives. These devices offer large storage capacity, but suffer from reliability concerns similar to those of conventional tape drives, and they tend to access data serially. You can't go in and read any file you'd like, play back from front to back, or back to front.

It is also worth noting that moving mechanical parts inherent in these devices can generate small but occasionally significant vibrational disturbances to a small satellite platform. Spinning devices like the tape spools or the metal platters on which modern hard drives store data add angular momentum, which can make attitude maneuvers more difficult. These forces are insignificantly small in your lap top, but the satellite lives in a much quieter environment. The stabilization force on some small satellites can be swamped by the force of the weight of a housefly.

Also, though some of these terrestrial computer components have been toughened for portable use, i.e., with laptop computers, they are still quite sensitive to vibration. Most difficult to handle is the harmonic, periodic loads of the launch. Just because a hard disk can withstand a drop from the tabletop to the floor does not mean it can stand the strong, prolonged, highly resonant vibrations of the launch vehicle and the tests designed to ensure the satellite will survive the launch.

Bubble memories were quite vogue for a few years in the `80s. Their main feature is that once data are written, they are not erased even if power is lost, unless they are written over intentionally. Bubble acts like random access memory (RAM) in a computer, but holds data even without power. But bubble memories have proven large, costly, heavy, and quite slow. Lacking acceptance in the terrestrial computer industry, they have not kept pace with other solid state devices. These technologies, particularly RAM, designed around semiconductor memories like those found in computers, offer many advantages. In fact, they are the only memory type that has flown on board a small satellite. While other technologies are being studied, particularly use of hard disks and optical disks, solid state memories have become so compact, reliable, and power conserving that the memory does not drive the spacecraft design, and the motivation to find an even better solution, if any even exist, is weak. So for now and the foreseeable future, RAM will be the choice of most small satellites and small rockets. They are the topic of the rest of this chapter.

Anatomy of a Solid State Memory System

The diagram below shows the three major components of a solid state memory system. The memory array, which provides the actual storage, is almost always expandable and configurable for particular applications. The interface section provides the path into the memory system; data are written and read from the memory through this path. The controller manages the memory array, keeping track of where the data are, what portions are used, and what portions are available for more data.

Semiconductor Memories

Three different types of semiconductor memories to consider for a spacecraft memory system are: dynamic random access memory (DRAM); flash electrically programmable read only memory (flash EPROM); and static random access memory (SRAM).

DRAM Storage

DRAMs are the semiconductor storage devices most commonly used in all computer systems for terrestrial applications like your Macintosh or PC. They have the advantage of the best density and lowest cost per bit. In the past, they were not selected much for space applications because they require frequent refreshing, which requires power. They have also had a low radiation tolerance. The power requirements have dropped a lot as manufacturers of laptops

Chapter 12—Memory Systems

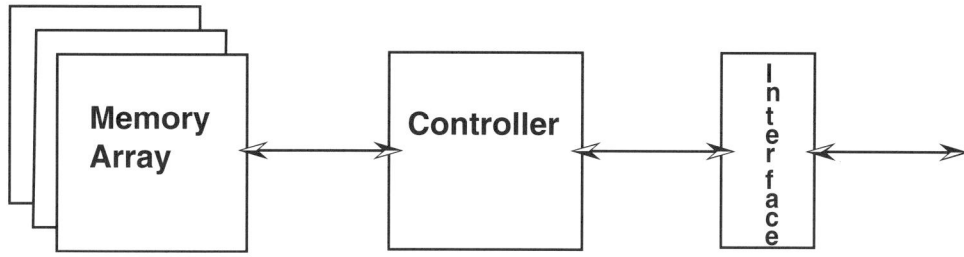

Solid State Memory System.

and other portable devices like cellular phones have innovated to combine low power with lots of fast, cheap memory. There has also been a serendipity in the radiation area. While this is not significant for terrestrial applications, radiation tolerance is vital for most LEO satellites. It has turned out that the processes necessary to get good manufacturing yields of very dense DRAM happens to yield quite radiation tolerant parts! Also, faster processors mean the satellite computer can check for and cleanse the memory of errors quite frequently, so that they can be caught and corrected. Thus, while DRAMs have not yet been popular for space applications, they will be flown in the near future.

EPROM Storage

Flash EPROM is a relatively new technology that allows unlimited reads. It can be electrically bulk-erased and reprogrammed for a limited number of times, approximately 10,000 write cycles. The main advantage of Flash EPROM is that once you write it, the data are permanently stored without requiring electrical power until you purposely erase them. Flash EPROM is dense, shows promise of good radiation tolerance, and can be completely shut down indefinitely for the minimum possible power requirement. For some applications, these devices promise to be an excellent match.

SRAM Technology

SRAM, which is today's semiconductor device of choice for most space flight applications, is the subject of the rest of this chapter.

Static RAMs offer low power consumption, high data transfer rates, reasonable density, and have a good commercial market base, which directly benefits their cost and also their density—how many bits of information you can store on each chip.

The current technology for SRAM is 1Mbit parts. Each part holds over a million bits of information. 4 Mbit parts are starting to become available. Historical trends indicate that the technology provides a quadrupling of capacity for a given part type every two or three years.

The rapid continuous advance of microelectronics technology makes it very advantageous to use the newest parts available for a memory system. This statement contradicts the approach taken on large satellite programs, where proven older technology is preferred over more advanced, but unproven, devices. The inability to accept risk has inhibited adaptation of large digital memories aboard conventional spacecraft.

Radiation Effects

One of the more significant differences in operating a memory system in a space environment compared to a terrestrial environment is the effect of radiation. Radiation has several effects with different consequences. Long-term exposure to radiation results in a gradual degradation of electronic components. This effect is referred to as total dose. Bias currents increase and thresholds change. At some point, the device stops working. This problem is solved with a two-part strategy: first, by selecting components with intrinsic hardness; and second, by shielding, which means packaging the components in a metallic enclosure to reduce their exposure.

Energetic ions are atoms and molecules missing a few electrons and hence carrying a net positive charge. These ions, present in the LEO environment, can imbed themselves into circuit element substrates

Chapter 12—Memory Systems

and cause some types of components to latch up. Essentially, the component suffers an internal short circuit. The strategy for dealing with latch-up is again two-part: careful selection backed up with "circuit breaker" protection circuits. Often, if the latched-up part has its power cut and restored, the part can be restored to normal.

More often an ion does not cause latch-up but instead causes a single event upset (SEU). SEUs occur when a circuit changes state, so that a stored 1 becomes a 0 or vice versa. If SEUs occur in the actual memory components, the data are corrupted; if they occur in the control circuitry, the memory management is disrupted.

Error Detection and Correction

SEUs in the memory array are handled by storing more bits of information than strictly necessary — essentially by storing redundant data. These allow reconstruction of the correct data even after some corruption.

The simplest example of this is parity. For every byte of data in memory, the system allocates and maintains an additional bit of parity information. If the total numbers of 1s in the data is an even number, the parity bit is set to 1. If the total number of 1s in the data is odd, the parity bit is set to 0. When the memory system is given a byte of new data to store, it calculates the current parity value for that byte. When the byte is read out of the system, it can be verified that the parity bit still agrees. This way, theoretically, if you suffer an SEU and a bit "flips," you'll know something's awry.

This scheme has numerous shortcomings. Parity is not a way to correct corruption. It only tells you if the data are good or bad, not how to fix them. Likewise, a byte with multiple bad bits is likely to be identified as good.

More sophisticated EDAC (error detection and correction) schemes store a number of parity bits for each data word, which allows correction of single bit errors and detection of many multiple bit errors. For example, a common scheme uses an algorithm known as 32-bit Hamming encoding. For every 32 bits of data, 6 additional bits of parity, or check, are generated and maintained.

Micro Space Craft

Imagine a situation where a memory array begins with correct, uncorrupted data. Over time, SEUs invalidate bits in the array. The memory now contains bad data. When the first few bits flip, the Hamming code can used to detect and correct the error, thus returning the memory to perfect fidelity. However, if nobody "looks" at the data and EDAC bits, these random bit flips start to add up. After a while, all of the data in the array may be corrupted, and therefore useless. No EDAC scheme can help you.

The solution is to periodically "scrub" or "traverse" the memory, going through each location, verifying the data, and correcting bad locations. How often scrubbing is required to avoid accumulating too many errors to affect a full correction is determined from a probability analysis of the frequency of bit errors. If you have an idea of the upset rate (which is a hard number to quantify) for the orbit of your mission and the RAM technology you are using, you can calculate the total error rate (the rate of accumulating errors which can't be scrubbed using the error correction schemes in place). This calculation is based upon the frequency of traversal, the number of memory bits, and the capabilities of the specific error detection and correction (EDAC) codes used.

For small memories, scrubbing is relatively easy. The microprocessor controlling the memory reads each block of data and uses the EDAC bits to fix occasional errors. However, as the memory grows in size, the task gets harder. Of course, twice the memory requires twice the speed in the processor to scrub the whole memory. Most larger memory systems have dedicated hardware specifically designed to speed the EDAC process.

Another important error correction function of the memory is to "map out" bad memory parts. Mapping out can eliminate potential errors due to hardware problems. If a memory chip, for example, should fail, the processor can eliminate that part of the memory from its address space. It just stops using the bad parts. Software sensitive to repetitive errors in specific addresses can be included for doing mapping out tasks.

Chapter 12—Memory Systems

Transfer Rate

The transfer rate tells how quickly data can be moved into and out of a memory system. A typical application might require the input rate to be fairly low, on the order of 10 Kbit (10,000 bits) per second, and the output rate to be higher, say 1 Mbit (1,000,000 bits) per second. This is typical because while data may trickle in slowly as the spacecraft orbits the earth, it must be downloaded rapidly during the brief period of time when it can communicate with the ground station.

Because static RAM is very fast, the transfer rate is not limited by the memory technology, but by the operating speed of the interface.

Interfaces

A wide variety of power and data interfaces to memory systems are available.

For the power interface, most systems can be configured to accept either unconditioned 28V (or any other voltage) or conditioned 5V supplies. From a systems point of view, the first option is easier; however, the second option can be slightly more efficient.

Data interfaces can be either serial or parallel. Serial interfaces are better from a cabling and configuration point of view; parallel interfaces allow faster data transfer rates and generally simpler support software.

Typical serial interfaces include simple asynchronous dedicated links, usually based upon the RS-232 standard, and multidrop serial links, for example, MODBUS, based upon the RS-485 standard and avionics standards like MIL-1553.

Parallel interfaces are most often specifically designed for a particular application.

Power Consumption

Static RAM uses very little power when quiescent; thus, the size of the memory array has little impact on the power consumption. You simply leave most of the memory in its quiescent state, and only activate the portions you need to read or write. What does drive power consumption is the data transfer rate and activity and the scrubbing rate for error correction.

Testing, Quality Assurance, and Reliability

One nice aspect of memory systems is that it is quite possible to do a thorough job of testing on the ground. Unlike systems that require the absence of gravity (flimsy deployables) or that a particular aspect of the space environment be completely characterized, memory systems are really only sensitive to launch loads and vibrations, radiation, and thermal conditions. A well-devised "shake and bake" test plan can provide high confidence in system integrity.

Radiation testing is more problematic, but it is a fairly mature, well understood art. Radiation testing an entire memory system would prove quite difficult; instead, parts are tested at the component level.

Electronic components are available in a wide range of screening levels. The basic parts are the same, but the higher screening levels are subjected to additional inspection and testing. Some of the levels, in ascending order of supposed quality and definitely cost, include commercial, Mil-883C Class B, and Class S. You typically find one order of magnitude increase in price for each step up the screening hierarchy. Note that you are NOT getting an order of magnitude increase in part quality for your money. You are getting more traceability and documentation and some assurance that the parts will not fail prematurely.

Typically, small satellite programs are cost-constrained. In many cases, small satellites are built with lower screening grades than would be used for a large conventional satellite. The thinking here is

Chapter 12—Memory Systems

that testing and part replacement is so easy in a small satellite that system level screening will weed out bad parts. Also, small satellites simply have less parts to fail, so they can accept lower reliability specs per part.

Memory systems by their very nature offer a high degree of graceful degradation. Generally, they do not fail at once, but rather individual SRAM chips fail. It can be very worthwhile to consider a lower cost part and accept the potentially higher failure rate, with the knowledge that this system can easily map out failed parts and continue to operate with only slightly reduced capacity.

Additional Processing Tasks

It is also worthwhile to consider that the memory controller in many systems is resident in the general purpose microprocessor that might have considerable extra computational power besides what's needed for the memory maintenance function. In some applications, this extra capacity can be used to advantage. Data could be prioritized so that important information is downlinked before mundane stuff, or data compression algorithms could be executed in background mode. Check with your memory vendor about support for these kinds of enhancements.

Vendors

Fairchild has built a number of space memory systems, some of which have been flown. Seakr Engineering builds a number of different memory systems, some with Shuttle flight heritage. Texas Instruments has recently announced a new memory system that looks very promising. AeroAstro has built memory systems for which detailed descriptions follow:

Case Study 1: ALEXIS/MOXE/EUVITA

AeroAstro has constructed three nearly identical memory units for space flight applications. A fourth system of higher density is now in development. The systems we have already delivered provide a benchmark for devices of this nature and demonstrate how quickly the technology is advancing compared to more modern units.

Micro Space Craft

These units were based upon 256 Kbit SRAMs made by Hitachi. The memory controller was based upon the Intel 8086 microprocessor, and the data interface was a custom parallel port. In some cases, redundant memory controllers were used.

Capacity: 1 Gbit (over 1,000,000,000 bits of useable data)
Interface: Custom 16 bit parallel
Transfer Rate: 1 Mbps max
Mass: 15kg (includes housing)
Volume: 18" x 10" x 7" (L x W x H)
Power: less than 3W at 5V regulated
Quality Level: MIL-883-B components, MIL-2000 workmanship
Radiation Tolerance: > 10k rad (Si)

This memory is shown in the photograph.

A 1 Gigabit memory using SRAM technology designed by AeroAstro in 1989 (cover removed).

Chapter 12—Memory Systems

Case Study 2: HETE/QUEN

This program, currently under development at AeroAstro, represents one generation of advance in memory technology. The system, based upon 1 Mbit SRAM technology, offers a huge increase in computational power for error correction, memory management, and additional processing tasks. The system controller is based upon two different microprocessors, one optimized for memory management and the other optimized for interface. Redundant memory controllers and data interfaces are available.

Capacity: 1 Gbit
Interface: Configurable to suit application, nominally serial (RS-422)
Transfer Rate: 2.5 Mbps; higher rates available
Mass: 5kg (includes housing)
Volume: 14" x 6" x 6" (L x W x H)
Power: less than 3W at 28V unregulated
Quality Level: Best commercial practices, higher levels available
Radiation Tolerance: > 10k rad (Si)

This design offers substantial mass and volume improvement over the previous generation. Both designs use more or less conventional packaging available off the shelf in military components. Higher densities are achievable through more advanced memory chip packaging. Exploiting denser packaging technologies now available can result in between a 4 and 10 time reduction in mass and volume compared with HETE/QUEN.

The Future?

SRAM is following the evolutionary path of bubble memory. It is not cost competitive with DRAM, and its power advantage has been eroded by advances in DRAM technology. DRAMs with 16 Mbits on a single chip are now available. This means that the entire spacecraft memory will fit on 56 die. Since several die can be fixed into one package, modern DRAMs allow an entire small spacecraft memory larger than the data which could be downlinked in a day to be placed on a portion of a circuit board. Even higher densities are possible, and advances in packaging multiple die may shrink the gigabit to a single "sugar cube" package of much less than 1 cubic inch (16 cc). Power consumption, not counting the spacecraft processor, which is necessary whether the memory is there or not, can be below 1 watt.

Until there are significant advances in the downlinking of data, and excepting some special case missions where many days or weeks might go by without a downlink opportunity (interplanetary missions could be one such case), memory simply is no longer a major issue in microsatellite design.

Chapter 13

Mechanisms:
The Nuts and Bolts of Small Satellites

Moving Parts

Just one typical integrated circuit, one of those tens or hundreds of little plastic bug-like creatures planted on their 32 steel legs on a typical circuit board inside your satellite or your home computer, has typically 100,000 active elements: simple switches (transistors); capacitors; resistors; logic gates; memory cells; each one capable of switching state millions of times a second. It is remarkably unremarkable that these innocuous, seemingly static things change state, in other words, do things, easily 1 billion times per second. A malfunction of a single gate, with a dimension of 10 millionths of a meter, can be the end of a multi-million dollar mission. You say you built in redundancy? Congratulations. Instead of sweating the failure of one in ten million elements, your reliable, redundant satellite is now vulnerable only if two failures occur out of twenty million active devices.

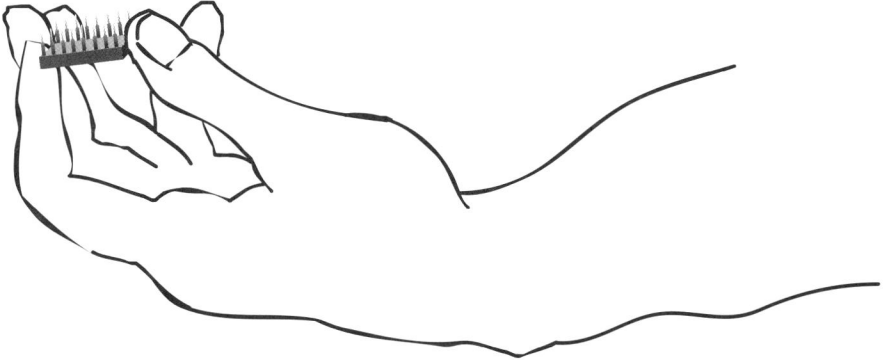

An IC chip with up to a million active elements — but no moving parts.

As far as I can tell, the steely-eyed macho satellite engineer has no problem accepting this risk. He'll stake his career on 10 million devices, each one of characteristic dimension 10 micron, all working fine at a state change rate of a billion per second, despite the reliability this implies for each one of those 99¢ DIPs (Dual In-line Pin) devices. But try telling Mr. or Ms. Cool that you want to deploy a boom with an antenna or magnetometer away from the satellite 5 or 10 feet. Or maybe you want to deploy the secondary mirror of a Casegranian telescope, or a solar collector or solar panel. Two moving parts, maybe three. You'll get two reactions: naked fear coupled with naked greed, for only by spending prodigiously can you assuage the fear. Something as "complex" as that could take millions of dollars to design, build, and test, and it will be the least reliable part of the whole space mission.

These moving parts are examples of deployables, which are things that are mechanically unfurled, moved out, or unfolded after the satellite gets on orbit. They are guaranteed to unnerve even the coolest designer. First comes denial.

"You really don't want to do that," the systems engineer explains, trying to remain rational.

Then anger.

"I thought you said you wanted a reliable, cheap satellite . . ." The designer imagines a black smudge on an otherwise perfect flight success record.

Then comes recrimination.

"I should have stuck with a simpler career, brain surgery or President of the United States."

Finally, acceptance.

"This is going to cost you big bucks."

Chapter 13—Mechanisms

Oh, the mighty dollar. Our ultimate weapon.

Hey, what's all the fuss? You've got a billion transistors on board, and all I'm asking for is one simple hinge. The difference is that none of those transistors physically moves. They are conductors, semiconductors, and insulators in and around which electrons flow. A deployable, like an antenna or sensor on a boom that extends on orbit, means physical motion and all that goes with it. Nasties like flexibility and poor alignment, leading to increased friction, jamming, and galling. It means force and restraint, momentum and potential deformation. Thermal inhomogeneity leading to dimensional instability. Vacuum welding and the problems associated with non-outgassing lubrication. Fine structures that can resonate and break. These are the lions and tigers and bears of spacecraft mechanisms. They have conspired with Murphy over the years to destroy billions of dollars in space missions.

Shall We Confront Our Fears?

On earth, we use deployables every day without fear or spending millions of dollars. Every morning I used to hop in my Buick, when I had a Buick, and switch on the radio. A whip antenna automatically, electrically deployed about 3 feet out of a 4" cylinder in the trunk, powered by a tiny electric motor. Tens of millions of cars have these, and they work great. If you break one off somehow, you are talking $179 to replace it. More than a McSalad, but not a big factor in a $20M spacecraft mission.

So why aren't we flying these marvels of reliability? This is where we space guys get Senator Proxmire's Golden Hammer Award. But not 'cause we want it. Let's follow the life history of that simple antenna deployment mechanism as if it were to be applied to spacecraft. For one thing, antenna pieces are made out of plastic that outgasses, meaning that in space they get rough and stiff, and the evaporated constituent materials could recondense elsewhere on the satellite, like on the surface of the optics. OK, so we replace that stuff, assuming there's a replacement with suitable properties.

Micro Space Craft

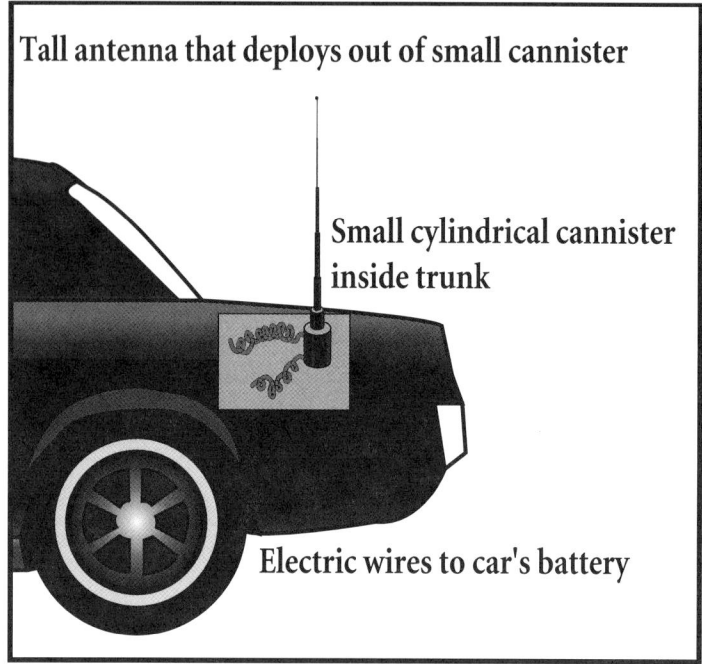

An earthbound deployable.

Next problem is that the Buick antenna has lots of metal-to-metal contact—surfaces that have to slide over one another. On the ground these are lubricated with oil or grease, but these materials can't withstand the space vacuum. They just evaporate away. Two metals touching each other generally do nothing on earth, but in the space vacuum, two metals pressed together can weld themselves together, their crystalline structures intermingling to form a bond. This technique is actually used by Peugeot to make cold welded aluminum bicycle frames, using a terrestrial vacuum chamber to join the various structural members, but you don't want your deployable antenna to try to weld itself together in orbit. So lubrication is a big issue.

The Buick antenna's motor depends on air circulation to cool it while it works, a feature not to be expected on a satellite, and it too is lubricated with space no-nos.

Chapter 13—Mechanisms

Space is a benign force environment, but getting there isn't. That Buick antenna is pretty tough, but can it survive 50 degrees below zero Fahrenheit (-40°C), and +175°F (+80°C)? Those are extremes that the mechanism might see on orbit, or at least on the way there. G-loads as high as 21 times the force of earth's gravity (21g) are encountered in some launches, and qualification tests might go to as many as 30g in vibration.

By the time you make all the design changes to use a motor that won't overheat and will operate without conventional lubricants; find metals that won't weld themselves together in vacuum and plastics that don't evaporate and that stay subtle and flexible over the full range of temperature; then beef it up so that everything can withstand the thermal and mechanical environments, you discover you can't actually build what the Buick engineers built.

As soon as you modify even innocuous small things, which the motor, lubricant, and basic structural material are not, you have a new, untested product. Which leads us to Dilemma #2 — that when you head off to build your own mousetrap, you will discover that the reliable and cheap Buick antenna actually took many years and many millions of dollars to get right. GM went through hundreds of prototypes and tested each of them for thousands of hours before fitting one on your car. Probably in the first year on the market, GM installed one million antennas, and got a million vehicle years of test data upon which they based a few additional design tweaks, which resulted in the nice reliable unit I got on my Buick. The reason it only costs $179 is that the development cost is spread over millions of cars every year for many years. If $50M were spent to build that deployable antenna and get it right, and it's used on 10 million cars a year for 5 years, that's $1 per car in R&D cost. You, on the other hand, want one deployable, and presumably you don't want to spend $50M to get it, nor do you have years to wait for it. There are companies that sell these spacecraft deployment devices, partly to spread the cost over many satellites. But they sell one or two or maybe five a year, so they still typically cost millions of dollars.

Micro Space Craft

In the real world, building a self-deploying antenna, or a solar panel on a hinge that simply folds out, or a rocket payload fairing that splits in two and disappears (in other words, leaves the vehicle) without leaving any pieces behind and without hitting the rocket or its payload, all turn out to be hard problems. We live in a society bathed in clever consumer devices, things like 500 MByte hard drives with discs the size of a silver dollar that can withstand the laptop computer they live in being dropped four feet onto the floor. We think these ought to cost $300, and they do because millions of them are made every year and their R&D costs are well distributed.

Like the days between autumn and the end of the year, your choices, when you need to deploy something aboard a small, inexpensive spacecraft, come down to a precious few:

1) Find a mechanism that has already flown and worked and try to finagle a good price for it;
2) find a terrestrial device that can work in space;
3) build something from scratch and try to beat the odds.

Small satellite developers play all three of these options. The weakest is the use of a previously qualified device. If it was built for a major mission, it is out of your price class. A few meters (10 feet) of deployable boom fully qualified for a major program can cost $2M. Or even $10M, which is more than the total budget of most small satellite development programs. Still, many of us have deployed things, and it's not a bad idea to call around and see what's on the proverbial shelf.

Terrestrial devices are tempting indeed. They work, and you don't pay any development cost for them, so they're cheap. But before you try to fly one, go through the following list of questions as a minimum:

- Can the terrestrial part withstand the temperature, vibration, and static load environment of my mission?

- Does it contain magnetic materials that can interfere with spacecraft instruments or create disturbance torques on orbit in the presence of earth's magnetic field?

Chapter 13—Mechanisms

- Will the materials it is made of outgas? Will they survive in space?

- Does it require lubrication? Can I find substitute lubricants and if so, will they work?

- Is vacuum welding possible?

- Does it require air to cool it? Will it sustain large temperature gradients without immersion in air that could cause warpage or failure?

- Does it contain any microelectronic parts that might not be radiation hard and that will therefore fry in the increased radiation environment of space?

- How reliable is it really? In space nobody is around to jiggle the knob or fuss with a balky connector. It works or it won't.

- What are its momentum characteristics? For instance, does it have rapidly spinning components that can cause the spacecraft to tumble when they spin up? Does it have angular momentum that can interfere with spacecraft stability?

- Flight safety, particularly if you launch on the Shuttle: Does it have any glass or other materials that can fracture and leave small pieces? What type of insulation is on the electric wires? Will it evaporate in vacuum and cause a short and potentially damage other components or maybe even drain the spacecraft battery through the short?

- How much power does it need? To a Buick, 100 watts is no real big deal, but that can be ten times the total power budget of a small satellite. Since most deployables only work for a few seconds, your major concern might only be instantaneous current capability of the electrical system.

- Can it be tested? If you have a device that can be adequately tested and demonstrated on the ground, that's a big plus. Beware of components too flimsy to work in the earth's 1g environment and at 1 atmosphere of pressure, since they are hard to prove workable at all.

- Finally, if you have to modify it to make it meet the space criteria, or to perform somehow better in space (for example, more angular precision in the way it's pointed once deployed), you should assume you are starting basically from scratch, since it will not work exactly like, nor will it have the reliability implicit in, the original device. It was the replacement of one simple wire that caused the recall of 50,000 Saturn automobiles. Remember, therefore, to budget for a significant test and qualification program.

In the words of a wise old technician I apprenticed under at a German wind tunnel: "Alles nicht so Einfach." Nothing is as simple [as we first think].

Finally, there's from-scratch development. All of the above disparaging comments notwithstanding, do-it-yourself can work. It has worked in the past, otherwise we wouldn't have the flight heritage designs we all try so hard to keep using today. Some pointers:

- Look at all the criteria listed above for using devices originally designed for terrestrial application.

- Make sure you test it over as wide a range of conditions as possible.

- Get a few uninvolved reviewers to look at your design, development, and test plans.

- Make it stronger, harder, bigger, and in general more overdesigned than you think you need to. You can always refine it in the next generation.

- Allow plenty of time (6 to 12 months) after the part is "finished" to make improvements and corrections.

- Design the spacecraft systems so that the failure of the home-brew part does not terminate the mission.

Chapter 13—Mechanisms

What's out there?

Following are a few deployment aids you might consider:

Pyrotechnic bolts. These favored classics have been around for many years. Space deployables have to withstand large launch loads, and then suddenly be free to move. One way to accomplish this feat is to use nice strong bolts fitted with explosive charges, often moving bolt cutters, to break them. They are reliable, but not particularly testable, because once tested, they can't be reused. You don't want to be in the same room when you initiate them. Pyrotechnic devices often impart high shock loads. They also require special circuitry to ensure that they absolutely will not fire prematurely. In other words, they are normally shorted to ensure that current doesn't flow through the initiator. Finally they aren't particularly cheap. A single bolt is $600 to $2500. The fact that they are probably the most commonly used actuator, despite all their disadvantages, speaks well for their advantages, which are reliability and a very solid, instantly removable structure.

Hot Wax Actuators. A relatively recent, though now flight-proven device, electrically heating and thus vaporizing wax captive within a small, stainless steel cylinder forces the motion of a piston that can be used to open a latch, release a coverplate or door, or push a moving mechanism on its way. Hot wax actuators are more expensive per unit than pyrotechnics, but they can be used and tested repeatedly. They have no special safety hazard, and they produce no shock loads. On the minus side, it can take 30 to 60 seconds to achieve a deployment, and the concomitant electrical energy requirement is high. Also unlike pyrotechnics, the hot wax actuator continues to draw current even after deployment is successfully achieved, so the spacecraft must include logic to terminate actuation. A final disadvantage: if anything goes wrong, Space News will probably report that the failure could have been anticipated because the deployable was held on only with wax! Hey, this actually happened to me. We live in a media ocean, and you've gotta think about swimming.

Melting wires, or I should say "vaporizing"? You can constrain a solar panel or antenna with a simple wire, part of which is heated

163

Micro Space Craft

electrically and vaporized, thus releasing the wire constraint. These are very inexpensive, safe, and shock-free. Some people worry about the vaporized metal and where on the spacecraft it might redeposit.

Sublimation. Hold two parts together with a plastic molded part made out of a material that outgasses 100%, meaning that it evaporates completely in space vacuum. When that happens, the two parts separate. This technique has been used to achieve very good thermal isolation by removing the mechanical supports of the isolated vessel. Again, redeposition of the sublimated material onto sensitive surfaces is a potential issue.

The deployment mechanisms themselves include:

Hinges. Hey, they're simple; they usually work.

Carpenter tape. An old AMSAT technique, antennas are made of metal carpenter tape rolled up. When released, the tape straightens out and voilá! An antenna. Most recently seen on Orbcomm, the technique has been used by amateurs for over 20 years.

Stacer. This word refers to a ribbon of metal rolled into a truncated cone. The metal ribbon wants to unwind into a long rod but is constrained by a pyrotechnic bolt. Stacers deploy with plenty of gusto. In fact, they can puncture arctic ice! Cheap but not hugely reliable, stacers rotate as they extend, making them less than the first choice if you need to tightly specify the deployed object's attitude. Stacers have been used to create low cost gravity gradient booms.

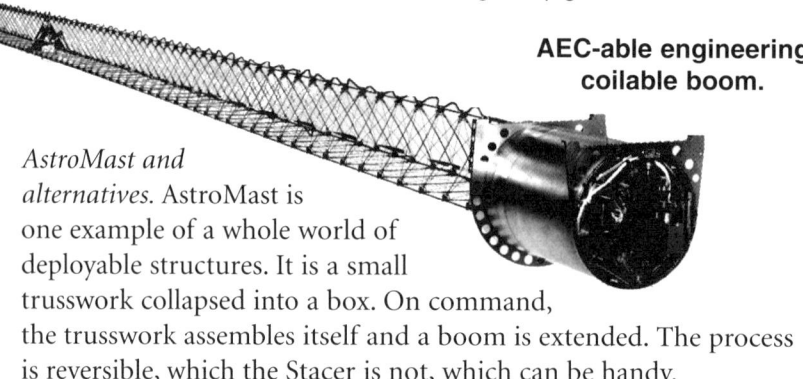

AEC-able engineering coilable boom.

AstroMast and alternatives. AstroMast is one example of a whole world of deployable structures. It is a small trusswork collapsed into a box. On command, the trusswork assembles itself and a boom is extended. The process is reversible, which the Stacer is not, which can be handy.

Chapter 13—Mechanisms

All kinds of shapes have been deployed: beams, flower petals, complex radiator geometry, multiply folded solar panels. Some of what has been produced is truly amazing, but that does NOT mean you want to try them. Most require years of development time and tens of millions of dollars to design, build, and qualify.

Testing

Space engineers appear obsessed with testing. Vacuum chambers, shake tables, anechoic chambers, radio ranges, all the high tech paraphernalia of preparation for life at 7 km/s, high above the atmosphere. Many parts of satellites are really quite easy to test. Radios could care less if they are in zero gravity or not, ditto (remember when people used to say ditto and there were machines that were named accordingly?) computers and batteries. Yeah, they need to be tested, particularly to ensure they can withstand the thermal and radiation environment, but the absence of gravity, hard to simulate on the ground anyway, is not a major issue for electronics.

But it sure is for mechanisms. Deploying a 10 kg (22 pound) mass at the end of a flimsy 10 meter (33 foot) boom might not even be possible on the ground. Many space booms can't support the mass they support in space in any orientation on earth. They can't push it upward, because their motors and springs are too weak. They can't push it downward because the momentum of the deployed mass causes the structure to fail, and they can't do it horizontally because the side loads are too great and will bend the boom. Big programs try to get around these problems in clever, albeit expensive, ways. For instance, they create very flat, 15 meter (50 foot) long stainless steel tables with thousands of tiny holes machined in them through which compressed air is pumped. These massive air hockey stadia are virtually frictionless horizontal surfaces and stuff gets deployed on them.

This is pretty interesting, but for low cost space work, useless. You can't afford one of these; you can't even afford to use one for a month, but it's interesting to know that if you could, you would. Just in case you think everybody in this business is hidebound and limited by their own imaginations, I know of one 0-g mechanism that was tested horizontally with helium balloons used to offset earth's gravity. Problem there is helium balloons have a lot of aerodynamic drag relative to their lifting force, hence they have trouble moving as fast as the mechanism and hence they produce a retarding force. In the end it worked, but not without significant fiddling around and pledges that we would never try that again!

The other side of the testing coin is that we really don't test things at all. Not compared with the commercial world. How many testing hours does that Buick antenna accumulate in its first year on the road? Say there's a million of them sold on all the GM cars, and each radio goes on and off twice a day, so that's two deployments, 2,000,000 a day, 365 days, for 730,000,000 deployments in the first year. You can build some failure statistics on that. How about testing a new drug? Take 1000 people and test it, or 10,000 people. That's a hugely expensive test program. But it's nothing compared with the number of Tylenol doses sold in a year, which is billions worldwide.

For most products, the field test is a final step in the testing process. Final testing might result in the occasional recall, but that is hard to avoid, given that real-world testing is millions of times more exacting, or at least more exhaustive, than the best lab test. Aerospace mechanisms don't get the opportunity to be tested in application. They have to work on their first time out.

A World of Mechanisms

Not all mechanisms are deployables. A very common mechanism is opening access doors. Sensitive optics or materials need to be kept in vacuum or otherwise isolated from the integration and launch environments. There are many, many sad stories of doors that refused to open on orbit, even after significant test programs on the ground. Some causes of failure have been launch vibration load, temperature extremes, and vacuum welding problems.

Other mechanisms include camera shutters and rotating carousels. Gyros and momentum wheels are also mechanisms, though they are often purchased as sealed units. The spacecraft designer doesn't need to be too concerned with their operation other than to be careful handling them. The bearings, lubricants, and assembly tech-

Chapter 13—Mechanisms

niques developed over many years to make these devices reliable shouldn't be taken too much for granted. A momentum wheel seems simple enough until you actually try to build one that will survive the launch and operate for many years on orbit.

The same can be said for computer hard disk drives. Problems flying these in space include:

- They make angular momentum that can interfere with the satellite stabilization.

- Their lifetime is relatively short compared with many space missions.

- They require an air environment to float the heads and for cooling, so they need to be in an airtight enclosure sealed for the duration of the mission.

- Though they can absorb shock, the are not made to withstand the extended, possibly harmonic vibration of launch.

- Their electronics are not radiation hard and hence they will not work in most orbits, even Low Earth Orbit (LEO), without major design.

A side comment on these devices is that modern satellites tend to be downlink, not memory, limited. For example, even with a 1 Mbit per second ground station, the typical 20 or so minutes of daily contact time to a LEO satellite from a single earth station amounts to only about 1 Gbit (10^9 bits) of downlink data a day, spread over several contacts. Thus, for many missions 1 Gbit of memory is plenty. This much memory, and more, can be built into less space using RAM than using a hard disk, avoiding a complex mechanism and a potential failure point.

You Turn Me On: Rock & Roll and Explosive Bolts

A certain coincidence has apparently gone unnoticed in the popular satellite press until now. (If any satellite press is popular; I mostly read it cause it's my job.) Satellites (the man-made kind) and rock

Micro Space Craft

and roll got started at the same time. As products of the '50s, they share certain commonalities, not the least of which is that people remain fascinated with them today for largely unfathomable reasons. Just what was so great about hula hoops and waitresses on roller skates at drive-ins? But in historical perspective, this coincidence explains why people who build satellites are obsessed with turning them on, a more modern, '60s kind of concept.

Think about the turning on part. A little satellite could weigh 50 kg (110 lb.). Maybe 10 of those kg (22 lbs.) are batteries. Sometimes it takes a month to get a satellite launched. And sometimes, as in the case of the Shuttle, you install the satellite six months before it gets into space where its solar panels can receive sunlight and recharge the batteries. Ever leave batteries in your favorite flashlight for six months and then turn it on? Zilch, right? Add the fact that NASA likes to launch satellites with their batteries discharged to make sure they don't do anything unexpected, and you begin to understand the problem.

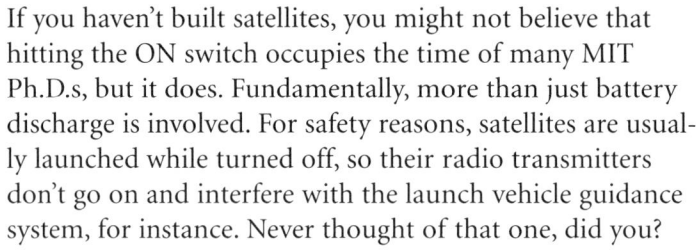

If you haven't built satellites, you might not believe that hitting the ON switch occupies the time of many MIT Ph.D.s, but it does. Fundamentally, more than just battery discharge is involved. For safety reasons, satellites are usually launched while turned off, so their radio transmitters don't go on and interfere with the launch vehicle guidance system, for instance. Never thought of that one, did you?

You know how you're not supposed to use the cellular phone when going through a road construction area where they're using explosives? Well, guess what a solid propellant rocket looks like? About 25,000 kg (55,000 lbs.) of blasting cap. If your downlink radio happens to resonate into that mass of plastic explosive, the launch crew is definitely NOT going to have a nice day.

So we're talking serious OFF here. Back to rock and roll. Besides OFF, the satellite is seriously ATTACHED to the launch vehicle. A ride into orbit on a typical launch vehicle has a lot in common with a real blood and guts rock and roll concert—serious noise, vibration, and physical abuse. The ride quality is 10 times worse than a

Chapter 13—Mechanisms

roller coaster Bart Simpson wouldn't touch. Test loads are often 20 times the force of gravity, so your 50 kg satellite has to be bolted down as if it weighed a ton. Then, after a few minutes similar to being bolted to one of KISS' big speakers on stage at LA's Fabulous Forum, your satellite is supposed to spring off lightly and amble quite civilly into its orbit.

One solution to these problems would be to build launch vehicles that aren't so sensitive to the occasional, highly unlikely, radio getting turned on and that don't shake their payloads within an inch of their lives. Until that becomes a reality, engineers spend a lot of their time Turning On, and to quote a favorite rock and roll cliche, Getting Off.

The de facto standard Getting Off technology is the Marman Ring, which is named after a real person who was not, contrary to often-seen spelling, a Mormon. He invented the method sketched below for making a really good mechanical joint, and then severing it simply, quickly, and symmetrically. By symmetrically I mean so that it doesn't come off tumbling (a tough problem for the satellite attitude control system), but straight. Axially. Probably a more generic term for this kind of separation design is "manacle ring."

The Marman, or manacle, ring separation system is actually composed of several parts. The rocket and the satellite use tapered rings, actually truncated cone shapes with a very flat, mated surface. The two mated surfaces are surrounded by the third ring, called the Marman band or manacle clamp band. This very springy metal belt

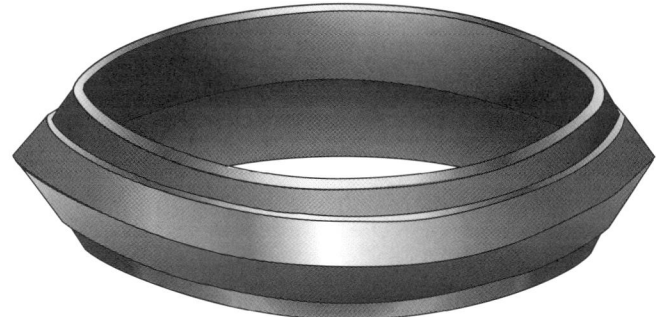

Marman or manacle ring.

169

Micro Space Craft

is bolted tightly around the joined satellite and launch vehicle rings. When the belt is correctly tightened, the two ring faces are held tightly together. If the belt is severed, the two surfaces are not attached at all. In practice, the belt itself is not severed, but rather the bolts that tightened it around the two rings. In fact, the band is often made of two halves bolted on both sides where the halves are joined.

Manacle ring separations mechanisms are used both for separating satellites and other payloads from vehicles, and for separating the stages and other parts of launch vehicles.

The bolts that hold the halves together are severed by pyrotechnic (explosive) bolt cutters. If either bolt is cut, the band, which is very springy and wants to straighten out flies off, flies off and the satellite is separated. Usually the bolts are fitted with two cutters each, so if any of the four cutters works, the satellite is separated. Many rings are fitted with springs that push the satellite away. Other times the launch vehicle backs out of the way. Some satellite carry propulsion systems that fire after the Marman Ring releases, and the satellite accelerates away.

Marman or manacle rings are to satellite separation as Mick Jagger is to rock and roll—sort of instant credibility. Nobody doubts Mick is the real thing. Tell any tough reviewer that you're using a manacle ring separation mechanism, and they're happy. Unlike Mick, who really is the real thing, the Marman Ring isn't constant as the Northern star, which Joni Mitchell once pointed out. Like her, over the years it's been reinterpreted many times, and the real physics behind its operation are somewhat subtle and beyond the current scope.

Make sure the Marman Ring you base your satellite's life on really is a Marman Ring and not just a circular something that oughta work. Potential problems include deformation of the rings under the stress of being tightened, which tends to held them together without the ring; vacuum welding (remember that grease is good for that 50s haircut, but really messy in space where it redeposits all over your clean satellite); and injuring the satellite as the band goes flying

Chapter 13—Mechanisms

off when it gets cut. Another problem is space junk—holding onto the pieces of the Marman band and the bolts so that they all stay with the rocket.

Bristol Aerospace of Winnipeg, Canada, is the leading supplier of manacle ring separations mechanisms (the real thing). Bristol has flown 1,1000 manacle ring mechanisms without a single flight failure.

There are other ways, of course. People have used the bolt cutting trick alone, for instance. Bolt the satellite to the launch vehicle and when you want to get off, cut the bolts. Trouble is, all the bolts have to cut, and all at the same time. But this does work. Slip rings cut by a circular charge, and lots of other ideas, are used in special circumstances. The Marman Ring isn't perfect; it's just well understood. It takes height, and that band is really moving when you cut it. It also requires explosive cutters, which means lots of safety qualifications. Still, that's where the technology is.

So you got off, but so far you didn't turn on, which is where aerospace engineering really goes paranoid. You really DON'T want it to turn on prematurely, but then it HAS to turn on. So half your brain is busy making sure it won't turn on, and the other half is making sure it will. This is like Spy vs. Spy, layer after layer, first an "inhibit" to keep it all off, no matter what, then another actuator to make sure it goes on, no matter what, right?

Typically, the action starts with a switch. Sometimes the switch is actuated when the manacle ring disappears. But that is sometimes too fast. If, for instance, the satellite wants to deploy solar panels when it gets turned on, you want to make sure that the rocket and the manacle ring are out of the way first, and that means letting a little time go by first. So OK, a switch and a timer. No big deal, right? But what if the timer doesn't go on? What about a second timer—maybe set the second one for a long time, like two weeks, and start it when you kiss the satellite goodbye on the launch pad. Why two weeks? What if it starts raining, they don't launch and they don't let you back at it? If it rains for five days and your timer is set for three, maybe the solar panels deploy inside the rocket. Fun.

Now about that switch. You're going to trust your baby to one little microswitch? Of course not. Put four in there, all in parallel. But then the safety people, who don't care if your satellite fails just so it doesn't hurt anything else when it does, say, "Gee, if any of those switches shorts prematurely, does the satellite go on and blow up the rocket?" So each switch has to go in series with another switch with a DIFFERENT actuation mechanism. Maybe it looks for sunlight or feels for the launch vehicle instead of for the Marman Ring.

Well, all of these arrangements could, and do, get tedious. You get the picture. With as many solutions as there are satellites and launch vehicles, you can bet that every satellite designer has agonized over this simple little conundrum. But this is good news for you. Don't reinvent the wheel. Go talk to your neighborhood satellite jock who is building her or his wheel, and paint it your own color.

The world is full of mechanisms: landing gear on aircraft, elevators and escalators, the animated characters at Disneyland. None of them have to survive a difficult environment like space, and all of them are the product of more testing than you can do on your pet mechanism in a lifetime. All of them failed at some time in their history. But, through a long process of refinement, the ones on airplanes and in amusement parks get fixed and improved and developed into the highly reliable devices we have today. Without that luxury, using existing, proven devices, overbuilding everything and making sure you can test at terrestrial conditions, along with following all of the guidelines for making mechanical structures for space is your best option. None of which is to say it can't be done.

We have deployed enormous structures in space, and even bigger ones are planned. Many of them have never failed to work perfectly, though did you ever wonder about that Apollo picture of the US flag planted on the Moon's surface, apparently waving in the breeze? Some breeze! — a deployment mechanism that didn't quite work as planned. Right in front of our eyes in one of the most famous of all space pictures, possibly one of the most famous images in the history of humanity, glares out at us a little lesson on deployable mechanisms. They sort of mostly work.

Chapter 14
Batteries Not Included

WHAT IS INCLUDED: useful information on electrical storage for small satellites sprinkled liberally upon a canvas of occasionally entertaining comments at best generally peripheral to the subject at hand…

Think about this simple phrase: Batteries Not Included. It's printed on the side of almost every package on sale at Toys-R-Us. Of course it means there are no batteries inside the item. It also means that you are going to need batteries before you use your toy. Since not you but that little 8-year-old monster of yours is going to unwrap the item and life immediately becomes nothing short of unbearable until the batteries somehow do mate up with the toy, you need to buy batteries. Not just any batteries—the right size and number. D? C? A? AA? AAA? 9V? Alkaline? NiCad? Transistor? Gold Top? How many? One, two, three, and some gadgets require 8 or 12! Used to be you could at least hypothetically live in a world without all these choices by moving to the Soviet Union, where the upper bound on battery types and sizes would closely approximate 1.

That option no longer available, you find yourself without recourse, standing in aisle 1b at Toys-R-Us, donning those reading glasses that make you look and feel like your mother-in-law, and scanning the fine print: Four D-size batteries required. You still don't know what kind of D-size, so you pick the cheapest, no, the second cheapest, lest the kid label him or herself Child of Misers, and off you go.

When you get home, your loving spouse tells you that you should have gotten NiCads (Nickel Cadmium batteries, aka rechargeable)

Micro Space Craft

because they're ecologically more desirable, or at least you should have snagged the disposable kind with reduced mercury. You slink off to bed, feeling a bit beaten by the whole experience, wherein you pick up your Newsweek only to find that Cadmium has replaced polyunsaturated fat as the bad guy of the year.

All that, dear reader, to run a little red fire engine fitted with a light and a siren whose sound, impinging on your eardrums already weakened from years of blasting Metallica on your (battery-powered) walkman and ghetto blaster (is English a great language or what!), is shorted directly to that little spot right behind your eyeballs that no aspirin tablet has ever been able to reach.

When You're Away

We know why walkmen, ghetto blasters, little red fire engines, and Nintendo games need batteries. They need electricity and normally you don't plug them into the wall. Satellites, whose batteries, believe it or not, are the subject of this chapter, have a taste for electricity too. As a homemaker/owner you probably think you own an infinity of extension cords (then why, you might ask yourself, can you never find one when you need one?). But you don't have enough extension cords to plug into your orbiting satellite! Even if you do, they'll get very messy as they wrap around the earth once every orbit.

Let the Sun Shine

Most satellites use photovoltaic (solar) cells arranged into neat panels to convert sunlight to electricity. This setup is nice when the sun is shining upon your panels, but rather inconvenient when the earth ambles between you and old Sol. Most small satellites end up in low earth orbit (LEO). Most LEO orbits are eclipsed from the sun by the earth for about 40 minutes out of each 100 minute orbital period. You need batteries if your panels are momentarily not pointing at the sun while your satellite does some attitude maneuvering or if you want to keep the satellite on during launch while it's enclosed in the rocket fairing, or after insertion in orbit before the solar panels are deployed. Some satellites have very high peak power loads. You turn on a big radio transmitter only when passing over your ground station in Lima, Ohio, which happens 20 minutes a day. Batteries

Chapter 14—Batteries Not Included

allow you to store the trickle of power from your solar cells all day long, and use that electric power in a big blast when you need it.

Have You Ever Had to Make Up Your Mind?

You will be glad to know that you probably face fewer choices in battery shopping for your satellite than for your kid's next weapon of mass discomfort. Spacecraft batteries don't sell nearly as well as consumer products. Research has proven that the annual cell output of the US' leading space battery manufacturer is less than the number of cells purchased in one busy hour at any single Toys-R-Us. Also, space application imposes lots of constraints on the battery design.

Ignore the few satellites that have been orbited without solar panels. These operate for only a few hours before the batteries expire or do not have any electrical systems. An example was the Echo passive reflector satellites flown in the '60s. And the moon. Satellite batteries need to be rechargeable. A very few make electricity from nuclear power, an option that would cost millions, except that it requires sign-off right on up to the President of the United States, which is safely relegated to the "not likely" file.

The only other thing in consumer products like NiCad (Nickel Cadmium, two fairly common metals that make up the electrodes of the battery), for some applications, for many years has been lead acid. Car batteries, which use a liquid, or aqueous, acid and lead as the electrodes, are of course rechargeable. They are the most commonly used lead acid batteries.

Besides NiCad and lead acid another rechargeable is available for satellite builders: Nickel Hydrogen (NiH_2). Battery manufacturers only recently discovered the reason NiH_2 has not been popular in small satellites. It's because there is no neat acronym like NiCad. Seriously, you think MacDonnell Douglas sold many planes before they went to MacDac? No way, José. They actually tried NiHy (for the battery, not the airplane company), but they were sued by a company that makes chocolate soda pop. All of this motivated development of a new battery called Nickel Metal Hydride, which has a nice acronym, NiMH. I always thought that stood for the National Institute of Mental Health. Wrong, but at least I remembered the acronym. Lithium batteries have been available to consumers for a long time, particularly for long life applications like watch batteries. The most advanced laptop computers and handheld video cameras use lithium ion rechargeable batteries. Eventually this technology will transition to satellites, where it will provide four to ten times the storage capacity of NiCad.

How Do They Stack Up?

Satellite batteries operate in a demanding environment. Our typical LEO satellite goes into the earth's penumbra (shade) 14 or 15 times a day over 5000 times a year, a huge amount of charging and discharging. To get good life for the satellite, we want them to withstand maybe 20,000 cycles. If you could get 20,000 cycle NiCads for the kid's toys, and recharged them every single day, your 8-year-old would be ready for retirement before the batteries would be!

Launch costs range from $1000 per kg to $100,000 per kg, so we also want the batteries to be light, small, and not to leak. You probably haven't seen a battery leak since you stopped buying those cheap paper ones that wrecked the seven-transistor radio you paid $6.95 (1963 dollars) for back at Lafayette Electronics. Note: That's roughly $55 today. Those little guys weren't such a bargain, were they! But because the batteries could be exposed to the vacuum environment, leakage is a problem in space.

Under certain overcharge conditions some batteries might need to vent gas, typically hydrogen, which could be hazardous. Under thoroughly abusive conditions, lead acid car batteries can vent gas, a preferable alternative to exploding.

NiH_2 batteries are the lightest option, storing about 50 watt-Hours of electricity per kg, compared to 35 WH per kg for NiCad and for lead acid. Besides its acronym, NiMH's attractiveness stems from its light weight, about 10% lighter than Nickel-Hydrogen.

Chapter 14—Batteries Not Included

This number is particularly important because, to maximize lifetime, satellite designers allow their batteries to routinely use only about 15% of their capacity.

For example, a pretty capable small satellite might have a 60 W continuous power requirement. Going into penumbra for 40 minutes results in 60 W x 40 minutes = 2400 W-minutes or 40 W-Hours of battery drain. That should be just over 1 kg of NiCad, and less than a kg of NiH_2, right? Wrong! Multiply those mass numbers by about 10 if you want lifetimes measured in months or years and not days or weeks. It is not unusual to require 15 kg of NiCad batteries on a spacecraft bus weighing less than 100 kg. That can make batteries the single largest consumer of your precious mass budget.

You Say Tomato, and I Say Tomahto…

Commercially available collection of six tiny cells.

In social circles hep to battery nomenclature, it is important to diferentiate between a "cell" and a "battery." When you buy a 9-volt battery (9V), it is really a collection of 6 tiny cells, each with 1.5V. A cell is a single cathode and anode, and typical single cell voltages are 1.2V for the rechargeables and 1.5V for disposables.

It is not really a good idea to plan to use these things at 1.2V, i.e., as single cells. To produce a peak power of a modest 100 W, you'd need 83 amperes, which would mean using cables about twice as thick as your automobile jumper cables.

Anyway, because most electronics need higher voltages, like 6 to 28 V, people tend to prefer higher voltages. To produce these voltages, we combine cells with smaller voltages into a group to create a battery with a higher voltage. When you shove 8 D cells into the old ghetto blaster, you are creating a 12 V battery out of 8 cells, each of which provides 1.5 V.

177

Micro Space Craft

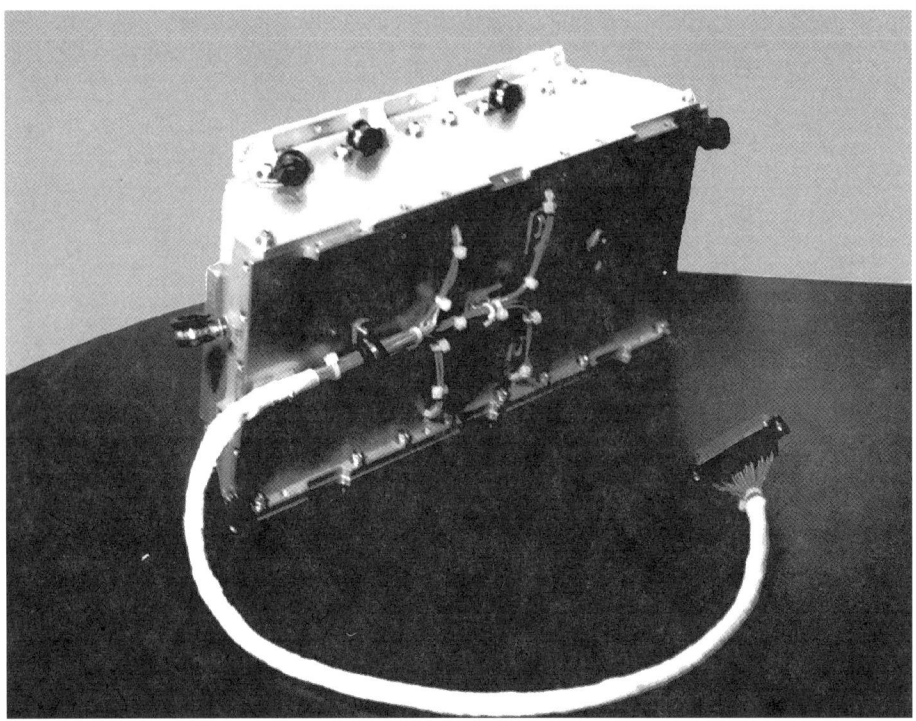

NiCad battery for the HETE satellite, containing 24 NiCad cells.

Now you know why they call them D-cells and C-cells that each put out just 1.5V, but 9-V batteries. And now you can make cocktail conversation about batteries even at highly chic aerospace gatherings without fear of embarrassment. In other chapters we addressed the motor vs. engine, perigee vs. perihelion, and satellite vs. spacecraft, but for now stick to batteries and stay away from orbits and engines. If asked about non-battery subjects, just excuse yourself and say you've got to run out to the car and use the cellular to check in with the babysitter.

William Safire, you might have noticed, I ain't. The point of the Cells vs. Batteries discourse is that to keep current (amperes) low, you need several cells in series. Most spacecraft use 28 V. Some use lower. But just as an example, a 28 V electrical bus requires about 24 cells, figuring that as they discharge they get down below 1.15 V per cell. The smallest NiH_2 cells are 30 Amp-Hour (AH). Configured into a 24 cell string, this is 720 Watt-Hours (WH).

Chapter 14—Batteries Not Included

Even assuming a depth of discharge of only 15%, that is 108 WH, appropriate for a spacecraft requiring about 165 W continuously. Most satellites carry at least two batteries consisting of two sets of 24 cells, thus providing enough for a 330 W continuous drain, which is about 10 times more than the most power-hungry small satellites. Indeed, two 24-cell strings of NiH_2 batteries would have a mass of 45 kg, possibly as much as the entire satellite. In practice, most small satellites depend on 4 AH or 6 AH cells, and 30 AH NiH_2 cells are just way too large for them.

NiH_2 batteries have one additional drawback for small satellites. The cells are round because the internal hydrogen gas pressure is high, typically about 65 atmospheres (950 psi). Rectangular prismatic NiCads pack much more densely. But to round out the picture, NiH_2 batteries also have excellent lifetime records and probably are capable of more charge and discharge cycles than any other space battery.

What about lead acid? It was bypassed for a long time for space application for several reasons. Clearly, flying a liquid acid like a car battery, including ports for gas venting overboard, looked like a bad idea for a satellite that needed to be tested in all attitudes. The acid would pour out! But lead acid batteries with a gelled acid do exist, like the gel formulation used in those leak-proof paper batteries that killed your seven-transistor radio back in '63. From that experience, you know lead acid can still leak. Most of today's pricier consumer disposable batteries are in metal cases to prevent leakage, as are the lead acids used for satellites. These so-called sealed lead acids have been used in a variety of field applications, including small satellites.

Lead acid cells are slightly heavier than NiCad compared to AH, but their voltage is slightly higher too, so in terms of power storage they are comparable. If you shop consumer batteries, you'll notice that NiCads don't last as long in your walkman as disposables, but they are also much lighter. Comparing a disposable C-cell with a rechargeable, for each cel the disposable or lead acid lasts longer. For each kg, they are comparable.

179

Qualify, Qualify, Qualify

There are no space-qualified lead acid batteries. You have to buy commercial units, test them, and take your chances. Lead acid batteries don't last as long as space-qualified NiCads, but they cost about $6 a cell vs. $2000 a cell for space-qualified NiCads. They come in cylindrical packages to provide a pressure sealed vessel to contain a small amount of generated gas, so nominally they do not vent.

Another low cost option is commercial-grade NiCads. If it worries you to use the same cells in your satellite as you do in your cordless screwdriver, then this approach is not for you, and neither is any lead acid variety, for that matter.

Commercial NiCads, like lead acids, run about $6 a cell, have power densities comparable to lead acid, and also come in cylindrical packages that look like the D-cells at Toys-R-Us.

This Year's Model

Last year's ecologist was pushing NiCad batteries for home use. Why throw out all that metal when you can buy a charger and recharge the same batteries hundreds of times? We can charge and discharge thousands of times in space applications where the depth of discharge is more limited.

This year Cadmium being introduced into the environment causes a lot of concern, not just from disposal issues, but because of the waste product of their manufacture. Our industry historically has believed that, as such a small part of the pollution picture, our designs shouldn't be torqued around by pollution effects only important on a much larger scale.

Nonetheless, every industry is now required to be ecologically responsible, and ultimately the future of NiCad technology could come into question. Lead acid, NiH_2, and NiMH are our available alternatives. Litigation is coming. In the meantime, in consumer application it is not clear whether the mercury-free disposables are ecologically more or less sound than NiCads. For right now, a ratio-

Chapter 14—Batteries Not Included

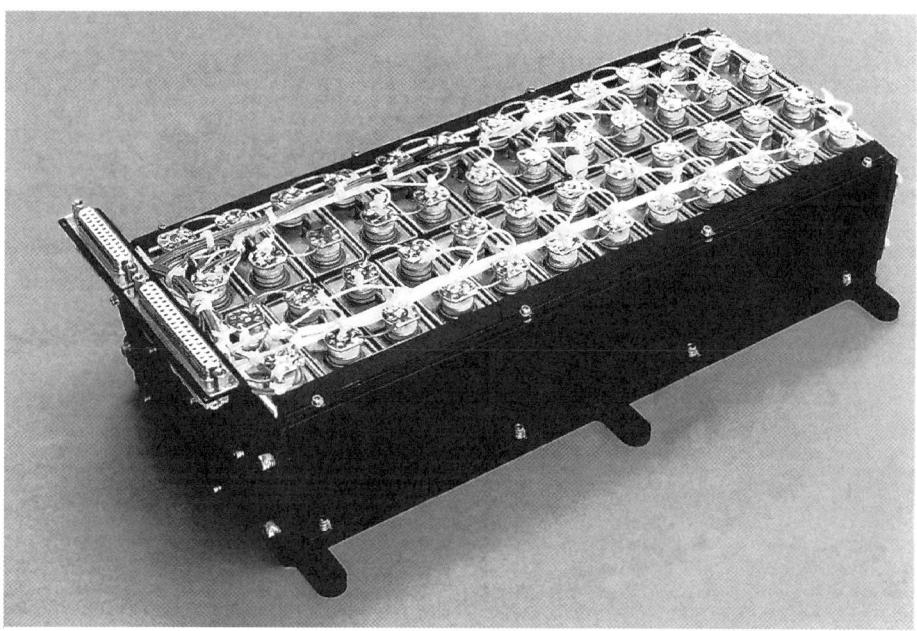

The Swedish Freja satellite used this space-qualified NiCad cell stack.

nal buying choice probably should depend more on economy than ecology.

Qualification II - the Sequel

Commercial cells have been used in space application for decades, but, like recycling, it has only recently become fashionable. Also like recycling, there are plenty of ways to do this incorrectly. Batteries must be carefully screened and matched. No matter how carefully you prequalify batteries, commercial units are more failure-prone and also do not survive as many charge and discharge cycles.

In addition, commercial units have a shorter shelf life because of diffusion of materials within the battery. This difficulty might be slowed by cold storage, but once any battery is placed in service, even if you don't charge and discharge it, it is aging. The biological clock ticks faster in commercial NiCads than in space-qualified units. Even if your mission lasts only a year, if the satellite sits in testing and in a warehouse, er, cleanroom, waiting for integration and launch, all that time counts against battery lifetime.

A space NiCad is a different animal than a consumer NiCad. The materials used in battery manufacture, and the manufacturing processes themselves, differ between commercial and space units, a contrast with many electronic components like ICs or discretes that might differ only in screening level. For this reason, application of commercial NiCads, while promising to save potentially hundreds of thousands of dollars, must be accompanied by prudent design and testing. Usually a larger number of commercial cells are used in order to ensure redundancy against a cell failure.

Solar Arrays for Small Spacecraft

The primary source of power for small LEO to geosynchronous spacecraft missions lasting from a few days to several years is solar arrays. They are an ideal source of power for spacecraft with long life and very high reliability. Several issues are key in determining the design of a solar array, including spacecraft configuration, required peak and average power levels, operating temperatures, shadowing, radiation environment, illumination or orientation, mission life, mass and area, cost, and risk.

Required Power Level and Mission Lifetime

The three key design considerations in sizing a solar array are mission lifetime; average power requirement, and orientation relative to the sun. Because of the degradation of solar cell performance over time due to radiation (refer to Radiation Degradation below), the solar array must be sized to meet the power requirements of the spacecraft at its End of Life (EOL), which, however, causes an oversupply of power to the spacecraft at the Beginning of Life (BOL). For missions with a lifetime of more than ten years, sources of power other than photovoltaics should be considered because of the degradation of the solar cells.

Chapter 14—Batteries Not Included

Operating Temperatures

A solar cell's performance and efficiency are affected greatly by temperature. The power characteristics of a cell are usually determined or quoted at 25 to 28 C. Each type of solar cell has a different temperature coefficient (% decline in efficiency / increase of 1 C). Silicon cells typically have a temperature coefficient of -0.46%/°C, while gallium-arsenide cells run at about -0.22%/°C.

Spacecraft Configuration

Because a single cell's power output is very low, cells in an array are usually arranged in strings. Cells are placed in series to reach the required voltage and in parallel to reach the required current. When a cell is not illuminated, it acts as an open circuit; therefore, the shadowing of a single cell in a string causes the loss of the entire string. This problem can be reduced by several methods: actively pointing and tracking solar arrays, using diodes, and arranging cells in a "ladder" network. On two- or three-axis stabilized spacecraft, tracking and pointing the solar array can minimize the effects of components, antennae or structure that might shadow the cells. Diodes may be used to bypass groups of solar cells in a string to help prevent damage to cells that are not illuminated. A ladder places several parallel strings in a series parallel network. In the case where a cell within a string is not illuminated, this ladder network offers the remaining cells an output current path, therefore reducing the degradation of the output.

Array Configuration

There are two types of solar array configurations: planar and concentrator. Either configuration can be body or panel (deployable) mounted. Planar arrays are by far the most often used arrangement for a solar array. The use of panel arrays is applicable to three- and sometimes two-axis stabilized spacecraft. Panel-mounted arrays may or may not be pointable to increase the power output of the array.

Micro Space Craft

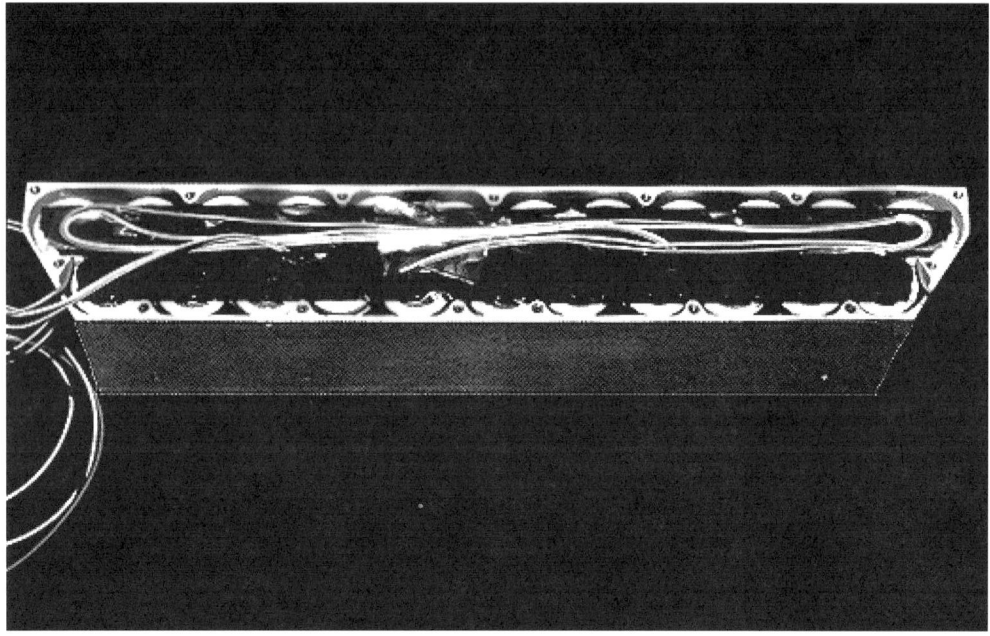

Battery box opened to show individual cells inside.

Body-mounted planar arrays, which are used on spinning or tumbling spacecraft, reduce the need for and expense of tracking panels. Using body-mounted panels produces a less efficient sun incidence angle that in turn reduces the array's efficiency. Body-mounted panels also operate at a higher temperature because they cannot radiate excess heat from the spacecraft into space, which also reduces the efficiency of the arrays.

Energy Conversion Efficiency

A cell's energy conversion efficiency is defined as the power output divided by the power input. The solar illumination intensity (1358 W/m2) is the input value for the power input to the planar array. For example, an array with an efficiency of 16% produces 217 W/m2.

Chapter 14—Batteries Not Included

Radiation Degradation

An inherent weakness of solar cells is their loss of effectiveness due to radiation in space. The degradation of energy conversion efficiency over the life of a solar array can be caused by several other factors such as thermal cycling, damage from particle impact, and material outgassing, to name a few. By far, however, radiation is the largest factor in the degradation of solar array performance. For a silicon array, the degradation in performance might be as much as 3.75% a year with 2.5% of that caused by radiation. For gallium-arsenide cells, the degradation is reduced to 2.75% a year, with 1.5% caused by radiation.

Illumination and Orientation

Orbital parameters such as the sun incident angles, eclipse periods, solar distance, and the concentration of solar energy influence a solar array's illumination and intensity. The sun incident angle is the angle between the normal angle to the solar array's surface and the sun vector. The output power of the solar array varies with the cosine of the sun incident angle, and is a very important consideration in designing the solar array to provide adequate power.

Chapter 15
Bring 'Em Up Clean

PLANNING TO BE the proud mom or dad of a budding little satellite? Dr. Sparck answers some new parents' most frequently asked questions on how to get Junior off to the healthy start she or he needs to ensure a long and healthy life on orbit.

Q: Dr., I've heard that developing young satellites need to be kept in a clean room 'till launch. What exactly is a clean room?

Dr. Sparck: There is no single definition of a clean room. What some parents consider clean is, to others, unacceptably dirty. Generally, a clean room has the components shown in the sketch below. All clean rooms have controlled access, including a staging area for donning special clean clothing. The idea is to minimize introduction of contaminants such as particles of hair, skin, lint, and dust into the clean environment. All tools, test equipment, and materials brought into the clean room must be thoroughly cleaned first. Special ceiling, wall, and floor coverings that do not generate dust also reduce the generation of contaminants, particularly drywall and paint dust.

An air circulating system introduces filtered air into the clean room and keeps the ambient pressure above the outside air pressure, preventing dirtier air from entering through doors and crevices. The air is moved rapidly in the clean room in a laminar stream from the ceiling toward the floor, so that newly generated particles are rapidly carried away.

Clean rooms are classified in terms of the number of particles per cubic FOOT of air. The most common level is class 10,000, meaning that there are on average 10,000 particles present in each cubic FOOT (about 1/30 cubic meter) of air. This level is about 100 times cleaner than a typical office environment. Some clean rooms used in aerospace fabrication achieve levels below class 1000. For some assembly operations, even cleaner environments are sometimes needed. Generally a clean chamber (glove box) or hooded tabletop is used, since achieving levels of class 10 to 100 is almost impossible if a human operator is present.

Q: *My home computer, walkman, and television all operate in a normal environment, and our house is no clean room! What's special about satellites that requires such particular care?*

Dr. Sparck: Young satellites most sensitive to dust are those with optics. Sensitive optical surfaces including special filters could be damaged by normal cleaning and hence must be covered or stored in a very clean environment. Van Der Waals and electrostatic attractive forces result in a gradual buildup of foreign material on lenses and coatings, but this buildup happens much more slowly in a clean environment. A special concern with UV (ultraviolet) optics is contamination by "organics" - organic molecules. Common hydrocarbons like waxes, lubricants, and cleaning solutions might be present in the air either as tiny droplets (aerosols) or as vapor that can condense on optical surfaces. The high UV absorbtivity of many hydrocarbons has created a huge market, such as skin cosmetic products that prevent sunburn. But if your spacecraft is sensitive to organics, special care needs to be taken in selecting paints, coatings, sealants, cleaning solutions, filter materials, clean room clothing and floor, ceiling and wall coverings that do not outgas UV-absorbing vapors.

To reduce weight, some spacecraft use extremely thin photovoltaic (solar cell) cover glass, too thin to withstand normal cleaning procedures. This glass must therefore be kept in a clean environment at all times. Historically, clean room environments have also been maintained to prevent small particles from potentially short circuiting fine electronic wiring.

Chapter 15—Bring 'Em Up Clean

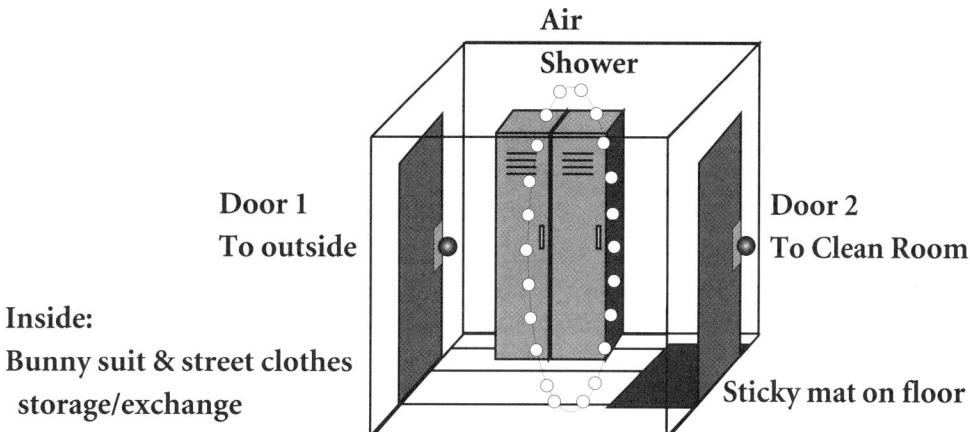

Entrance to clean room.

Q: But Doctor, some satellites are built in people's basements, garages, and living rooms. How do I know what's right for my family?

Dr. Sparck: We all want what's best for our children, and it's true that dust, dirt, and debris can cause failures and performance degradation. But keeping a child in a completely clean environment isn't the answer for every parent. Clean room upbringing is expensive. Besides the cost of building and maintaining the clean space, satellite care is much less efficient. Every tool that goes into the clean room must be swabbed, usually with alcohol using lint-free cloth.

The few minutes required to suit up each time you enter the clean room add up. Often workers find themselves changing into and out of clean room hats, masks, gloves, overalls, and boots five or ten times a day. This ritual is tiring and time consuming. There is a temptation to make do with inadequate or inappropriate tools rather than go through the changing process for the 11th time during a long shift. But many tools simply can't be used inside the clean room, including most heater and refrigeration units, which require purchase of special gear.

No little satellite lives its whole life in one room. Eventually it has to go to vibration and thermal vacuum testing, range tests for the

Micro Space Craft

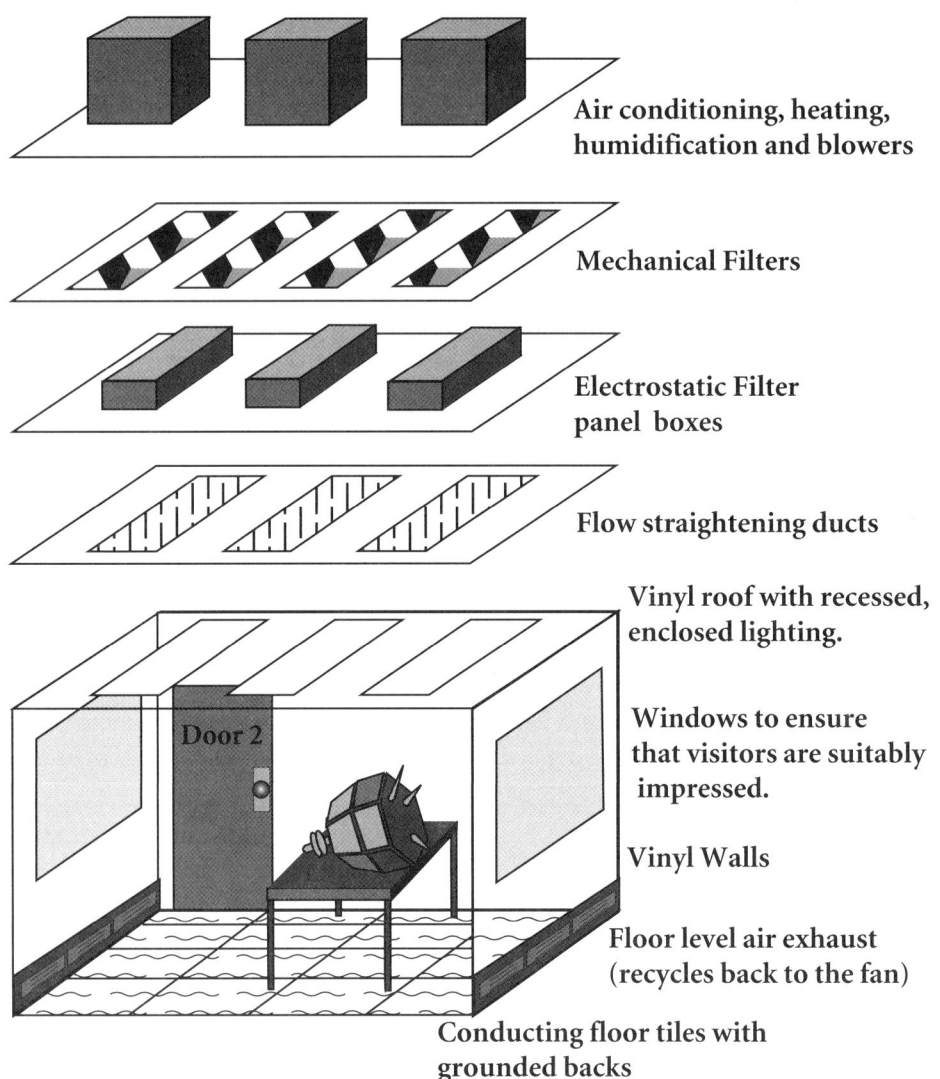

A diagram of the Complete Clean Room.

radios and antennas, and the launch site. If you bring up the child to thrive only in a clean room environment, all of these operations are slower and more expensive.

On the plus side, the clean room is a barrier to more than just dust. When your cousin, Butterfingers Bobby (known as Bull at the family reunions for the time he accidentally kicked over both flaming

Chapter 15—Bring 'Em Up Clean

barbecues, demonstrating once AGAIN that elegant place kick of his that won the Big Game down at State U 15 or was it 17 years ago?), shows up to bounce the new baby on his knee, you'll be glad to have the airlock between him and your pride and joy.

Clean rooms are the Prams of satellite rearing. Politicians, reporters, financiers, and traditionally minded engineers believe a gleaming clean room to be the sine qua non of professional satellite development. When they see it, they'll believe you've really arrived!

Many small satellites have no optics at all, or simply use covers to protect the optics. No small satellites have used photovoltaics too fragile for cleaning with a lint-free cloth and alcohol. The hazard of dust contamination of electronics is remote, and can be controlled in many other ways. Cost-conscious parents can raise a fine satellite without the benefit of a clean room.

Q: What are these other steps we can take to keep the satellite clean outside the clean room environment?

Dr. Sparck: The most important step is care. Serious contamination occurs both in and out of clean rooms through human error and facilities problems like leaking ceilings, insect invasions, spilling liquids, cutting wires and letting the shreds fall onto the satellite. The cleanest clean room isn't going to help you if you shear off a bolt head and it and its shreds fall into the electronics box!

Most electronics can be cleaned. Periodic swabbing with approved cleaners, for example, alcohol using lint-free cloth, keeps dust and debris down. The same is true of the spacecraft structure. Keep it clean. Use only cleaning, lubricating, and coating materials known to be approved for spacecraft, which usually means they have no organic contaminants, don't produce dust, and don't outgas. I know the can says it's made for rockets, but save your WD-40 for squeaky hinges on the Dodge, not on your mini-Delta.

Conformal coating is an option. It's a thick, plastic-y material that you paint onto circuit boards, but make sure you clean them first. Special cleaners are available that will ensure the boards are clean enough to guarantee adhesion of the coating. Conformal coating provides an impervious coating and protects your components. Most conformal coats are so tough, they provide some mechanical support to components that might be in danger of being overly stressed in vibration testing and launch. But don't count on it.

Something else you can count on, too, is that conformal coats are so tough that you can't get them off when you need to. If you fry a component on your board, no guarantee says you'll be able to cut through the coating and replace it. If you attempt to slice the coating off, you could damage other components and circuit board traces in the process. Despite this irksome drawback, conformal coating is used in a large fraction of satellite programs. One last caveat, however. The conformal coating only protects the surfaces you coat, and that can never be every surface. It does not eliminate the requirement for care in handling and working around satellites.

The oldest and most reliable means of ensuring cleanliness, particularly in sensitive optics, is covers. Covers are more effective than clean rooms because there is no air circulation to introduce new dust particles onto surfaces. Also, they can protect against the occasional dropped screwdriver, sneeze, and flying wire shaving. But covers have their drawbacks too. First, you have to remove them before flight. Murphy has reared his ugly head several times on this basic precept so I will not, for the sake of the memory of the dead and the careers of the living, divulge any specifics. Covers built to remove themselves in flight using mechanical actuators can fail. At any rate, the cover needs to come off, either for a test or in orbit. If the rest of the satellite resembles Pigpen, the optics do too, thanks to outgassing and diffusive motion.

The combination of careful handling, cleanliness, and using covers is the most common cleanliness system in small satellite practice today.

Q: Besides dust and dirt, what else do I have to be concerned about?

Dr. Sparck: The Number One hazard to the developing young satellite is not dust and dirt. The glamor of the clean room environment with the gleaming satellite surrounded by people in white clean

Chapter 15—Bring 'Em Up Clean

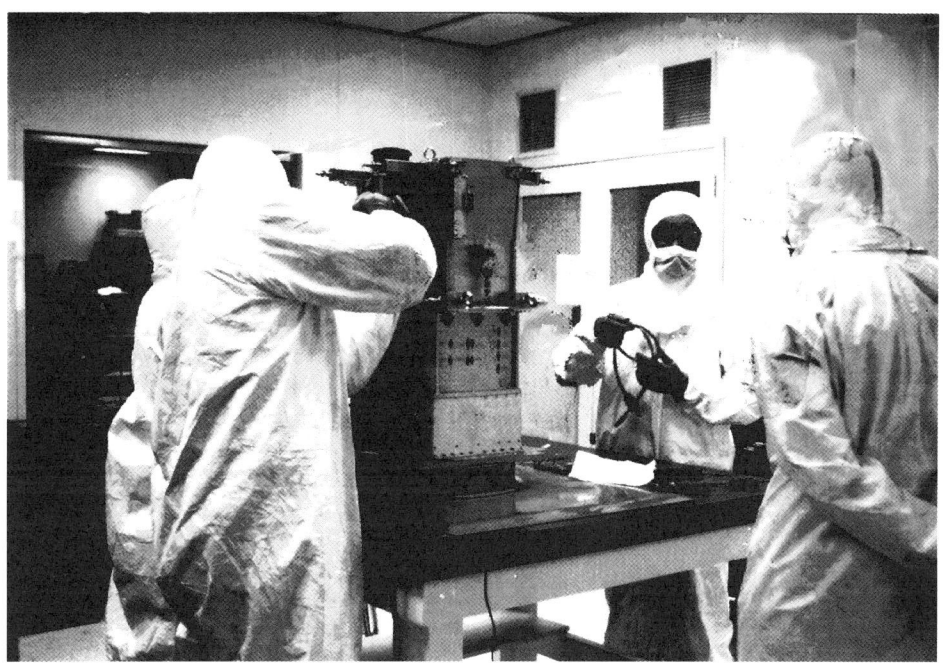

A clean room that Mother would approve.

room garb has brought cleanliness into the public awareness. The real killer is static electricity or, more exactly, electrostatic discharge (ESD). Particularly in dry environments such as heated indoor spaces, charge potentials of many tens of thousands of volts can build up between objects. Discharge of this voltage can destroy sensitive electronics. Maybe worse, it can cause latent damage unapparent for many days or even months of operation; then, without any other abuse, the component mysteriously fails, possibly during on-orbit operation where the mission might be compromised or lost.

Q: But what can I do about static—ESD?

Dr. Sparck: A lot! We know a lot about ESD, and it can be completely prevented. The first line of defense is to properly condition the air in areas where the spacecraft and its electronic components are handled, assembled, tested, and stored. Relative humidity should be controlled to between 45% and 65%. While higher humidity inhibits ESD, it can also cause condensation and possibly corrosion.

Micro Space Craft

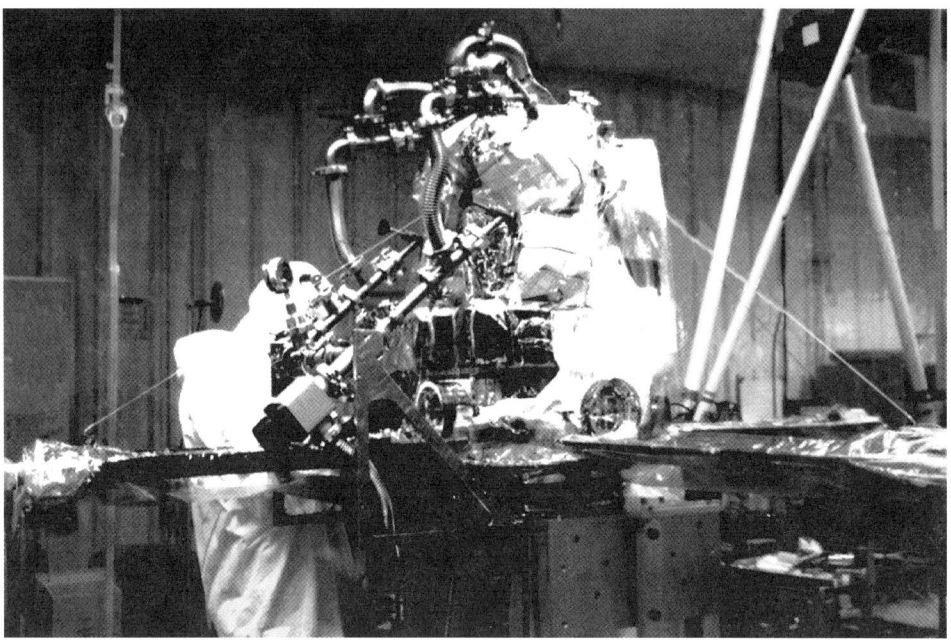

ALEXES satellite being mated to Pegasus rocket is protected by a portable clean room created with clear plastic curtains.

But humidity control is just a start. ESD-sensitive components must be stored in conductive plastic (pink poly is one common name, but the heavy gray-toned plastic ICs come wrapped in is also conductive). Just because plastic is pink or gray does NOT mean it is conductive. Make sure your plastic is conductive, and also that it does not outgas or cause organic contamination. Many packing foam materials are available that are antistatic and non-outgassing. Use ONLY these.

Absolutely critical is human handling of components, boards, and systems, for example, electronics chassis. You must be grounded at all times when handling electronics. Grounding is usually accomplished by wearing a flexible metal bracelet with a wire attached to a grounding post. Commercial products are available that show a green light when proper ground path is provided, red when the strap is not grounded. Rooms where electronics are handled should be static-protected. This means the floor should be electrically conductive AND GROUNDED.

Chapter 15—Bring 'Em Up Clean

Both vinyl and carpet floor coverings are available in conductive materials. They should be laid down over a conductive substrate, typically a copper sheet that itself is grounded. ESD furniture should be used. These are chairs and tables built of conductive material that are grounded to the floor at all times. They are surfaced with materials that do not tend to create charge when you or your clothes brush against them.

Beware of all cleaning procedures. They usually involve a process of rubbing a gas (like an air jet) or a cloth against the material to be cleaned. Like rubbing a party balloon on your hair to build up charge to make it stick to the wall or ceiling electrostatically, rubbing electronics with a cloth or a stream of air can build up tens of kilovolts of potential during cleaning. Air supplies and their canisters and nozzles must be grounded and only conductive cloths should be used for cleaning.

With these simple steps—humidity control, use of appropriate packaging materials, ground straps, conductive floor covering, care in cleaning and use of ESD protected furniture—you can protect your electronics from ESD. Remember, ESD damage is insidious. It can weaken a part without causing it to fail immediately. And, ESD-induced degradation cannot be detected.

Q: OK, we know to be careful about cleanliness and about ESD. Anything else?

Dr. Sparck: Only a few things, but they're important. Don't forget all this discipline when your satellite goes out to play. Wrap it in clean, non-outgassing anti-static materials. Make sure the satellite is grounded to shipping containers, and that they are vented only through adequate filters. Young satellites are prone to damage through mishandling in transportation. Shipping containers should include shock mounting and so-called shockwatch indicators. These record whether the maximum g-loading (acceleration) applied to the shipment exceeds a preset value. If the shockwatch indicator is tripped when you receive the package, the satellite should be thoroughly tested for possible damage. If it is not tripped, things are probably OK.

Micro Space Craft

The worst environment for a satellite is the salty air typical of many of the world's launch sites, including Cape Canaveral (Florida), Kourou (French Guyana), Barking Sands (Hawaii), Wallops Island (Virginia), Kagoshima (Japan), and Andøya (Norway). A responsible parent reviews the accommodations the satellite will have at integration and at launch to make sure exposure to the corrosive salt environment is minimized.

Personally, I believe most satellite damage isn't the result of any of these sophisticated hazards. How many of the cameras, watches, radios, and tape players you've owned have broken from vibration, static discharge, corrosion, what we would think of as intrinsic failures? Compared these with extrinsic failures caused by, say, dropping them overboard from your two-person sailboat, wearing them out in the rain, running them over with your car while pulling into the garage, or putting the batteries in backwards. These devices seem more reliable than we are!

When handling the satellite, use caution, allow a minimum number of people present at one time (because they distract each other), follow procedures, and NEVER rush or work to exhaustion.

The satellite is no more reliable than you are in handling it. If you don't make a big mistake, if you don't drop it, connect the plus lead to the minus pole and vice versa, drop things into it, spill things on it, or just bump into it (ever see a solar panel cover glass crack itself?), you've eliminated the major causes of failure. Use of protective covers is, last time I checked, not against the law either.

If any of these things do happen, and at least one of them will (probably many more than one in the course of development of a small satellite), your best defense is honesty. Report the event to the project team, analyze what damage might have occurred, and test to see if in fact there has been damage. An extra disassembly and cleaning is a preferable alternative to flying a satellite with a nut and washer loose inside it somewhere, waiting to lodge in a mechanism or short an electrical lead.

Chapter 16
Satellite Clusters

For almost four years, I have been recording the fine, but important, distinction between actual satellite programs and talk about satellite programs. Politicians, bureaucrats, lawyers, and the general public don't know or much care about the difference, but hardware building types don't thrive on air and paper. This need for nutrition in part explains why young, talented Americans are giving up engineering for more productive endeavors including skateboarding, legal work, management consulting, and entertainment.

The latest manifestation of the satellite equivalent of vaporware is the satellite cluster. While they didn't invent it, Motorola gets the credit for moving the topic into the Wall Street Journal with its announcement of a plan to place 77 small satellites in low earth orbit to provide global cellular telephone coverage. Several other firms are proposing clusters of small satellites to provide new and/or improved communications services.

Why a cluster, or a duck, for that matter? Way back before the big Bon Jovi disappointment, man envisioned two ways to communicate with satellites. One was to put individual, large, complex satellites into very high, 24,000-mile orbits that appear stationary in the sky, known as geosynchronous or GEO orbit. The other was to put many smaller satellites in low earth orbit. Because satellites in low orbit circle the globe about every 100 minutes, they move in and out of sight quickly. (Refer to the articles on Orbit Mechanics for a thoroughly weird but largely accurate treatment of satellite orbits.)

Micro Space Craft

Those who remember Telstar know the problem. It worked great for about five minutes.

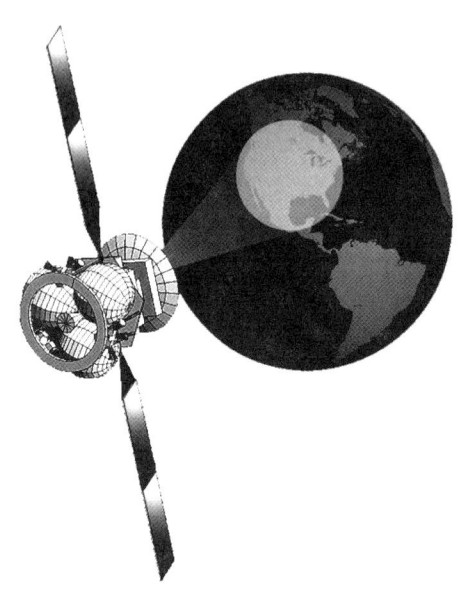

Large communications satellite at GEO.

The way to get over the disappearing satellite problem is to fly a cluster of small satellites. Each one provides a few minutes of service to any one particular user. Then its role is taken over by another cluster member. An earth-bound analogy is the cellular phone. As you drive from Costa Mesa to Mesa Blanca or from Cerritos to Cerillos (or, for that matter, Pawtucket to Pawkatuck), you move from cell to cell. You never realize that your phone is switching its attention from one cell to another. A cluster of satellites is like a cellular system except that you, the user, are relatively stationary, and the cells are circling the globe at four miles a second.

"So what?" you say. Hey, I'm getting to that. Large satellites in GEO have a few disadvantages. Take, for instance, the point of view of the poor soul trying to finance one of these collosaltrons. You go to the bank and tell them you need, say, $200M in small bills, all of which may, if some well-meaning technician has a bad day in the clean room and leaves a rag in a fuel line, be essentially destroyed and their remains sunk to the bottom of some nameless and deep ocean trench. (This happened.) The thought of such risk irks those buttoned-down bankers, a notably nervous lot, of late, who then send our friend the satellite entrepreneur to the insurance companies. They don't like betting all their marbles on one launch either, and life gets expensive, varying to impossible.

The small satellites, destined to become part of a large constellation of their brethren, are put in orbit one or a few at a time, thus requiring several launches and an explanation of why launch com-

Chapter 16—Satellite Clusters

panies think clusters are The Way to Go. A launch accident or a failed satellite is a 5% perturbation on the program funding and hence no big deal.

Another advantage often touted is that low earth orbiting (LEO) satellites are, well, low down (don't really need an ivy education to get that far, eh?). Hence, you don't need that 8' diameter status symbol parked in the back yard to tell your neighbors you've got a satellite receiver. You can reach the satellite with an antenna about as complex as the one on your FM transistor radio (like that nostalgic imagery?). To be fair to Mr. and Ms. Big Satellite, this assertion is not rigorously correct. Big satellites have big antennas, and they can use spot beams to give very good service to small antennas.

Ultimately, the little guys in the white hats win because of another subtle point. Being 20 or 30 or 77 in number, they can be close to everyone on earth simultaneously. That would be more than just a lot of spot beams on a big satellite. It would be impossible. A single geosynchronous satellite doesn't see the whole earth. You'd need about three to do it, assuming you could put a lot of agile antennas to work on each one, not an inexpensive proposition. The teeming multitudes at the North and South Poles (Mrs. Claus, the elves, and their nuclear submarines parked under the ice)

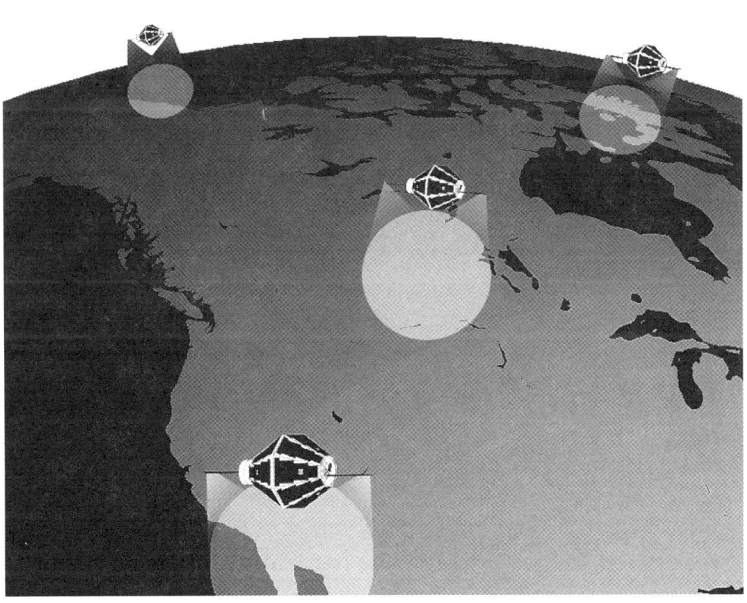

LEO satellites.

would still be out of range unless they use those big dishes that don't transport well on sleighs (tangle in the bells) or on submarines (attract sharks and rust a lot).

When you're alone in the car with them and the radio is blaring to avoid eavesdroping, the old-timers (ten years older than I am) will tell you a story about the engineers at Bell Labs who pioneered low earth orbit communications links in the 60s. They carried the day technically, but politics prevailed. The federal government threw its weight behind large geosynchronous systems as a way of justifying the large rockets needed so that our guys with the white scarves could beat the Russian guys with the red scarves to the moon. Of course that lunar base has been a longstanding cornerstone of our nation's defense, so nowadays it's considered well worth opposing compelling reasons to use small, low earth orbiting satellite clusters. Heresy, of course. Our government would never meddle in commerce and technology, but I wanted to represent the views held by a few thousand old satellite folks.

Well, one way or another, big geosynchronous satellites became the mainstream and weze guyz became a technical curiosity. Counting up the clusters already announced, we are looking at 175 satellites totalling over $3B in project funding. (And they said small satellites were just a hobby!)

What Are These Satellites Doing To Keep Busy Up There?

Starsys, Orbcomm, and Smart Car are aiming for the low cost end of the market with message forwarding services. Orbcomm and Smart Car would like to see a small box in every car's glove box with a connection to the radio antenna that every car already has. Break down by the side of the road? Of course not! Cars don't break down by the side of the road any more than people die in cemeteries. They break down in the middle of the road and then, except during snowstorms in Washington, DC, move to the side of the road so as not to get hit or just to stay out of the way. No matter. Smart Car or Orbcomm will allow me to set the button to message #11: "Broke Down By Side of Road," push Enter, and my message pops up at

Chapter 16—Satellite Clusters

some AAA or police headquarters along with the latitude and longitude of my rusted hulk. (The car, that is.)

Or I'm out hiking and break an ankle wrassling with a bear on some rocks. No problem! Set the little box in my back pack to message #83: "Broken Ankle, Road Rash, and Bear Claws," and the park service gets the message along with my position and I wake up in a helicopter with a paramedic casually nursing my wounds and furiously attempting to verify my VISA credit card limit. With ground terminals selling in the range of $50, the Orbcomm/Smart Car/Starnet systems intend to service millions of users with a basic, reliable, immediate means to send coded messages worldwide.

The simplest cluster configurations: to carry a message from a remote location, the nearest satellite of a cluster receives your message and carries it immediately for downlink to the addressee when it passes overhead.

Well, almost worldwide. Technology and politics clash again. The satellites are glad to provide worldwide service. Governments aren't. World radio regulations, devised for geosynchronous satellites with their narrow spot beams and unchanging coverage, are not ready for a satellite that covers the whole earth. It came up for the first time at WARC (World Radio Allocation Conference) in 1992. So-called mobile satellite (I have never understood this moniker. Are their immobile satellites?) service will probably keep lawyers and politicians busy for decades, while the world's satellite builders and markets wait for an elusive resolution.

Once again I digress. Next up the food chain from the low cost message and position relay satellites is Ellipsat, so named because the satellites are in a slightly elliptical orbit to maximize their time over North America. This is a concession to WARC since the service will

201

only be approved, at least initially, in the U.S. Though the satellites will cover the earth, they'll only operate over the U.S., maybe Canada. Kind of a waste, but they can spend their minutes over New Zealand recharging batteries. With plenty of time for R&R, the satellites can be built lighter and cheaper so it's not a total waste. A big part of OrbComm's pitch is focused on how the system is configured so that it is impossible for the satellites to carry messages across national boundaries. A complex system of earth station "gateways" and on-board software is designed to emasculate the power of their global cluster, a concession to the inability of politicians to catch up with technology.

Ellipsat, shown below, allows you to pick up your cellular phone anywhere. Maybe you're not in range of a cell, for example, you're rowing your boat from Block Island to Narraganset dragging most of your mast and sail in the water behind you. Switch your phone to overdrive (satellite) mode to reach one of the 24 Ellipsat satellites as it passes overhead. It relays your call through the Cranston East cell just as if you were in Cranston (you can always dream). The system augments the existing cellular network by getting you into it even when you're out of its normal coverage range. Since it is only a link into the existing phone network, it is relatively inexpensive and simple— if you can call $400M and 24 satellites inexpensive and simple.

More capable clusters have enough satellites to give continuous service linking users to the land-based telephone system.

To put inexpensive and simple in proper perspective, there's Motorola's $2.6B Iridium. Shown

Chapter 16—Satellite Clusters

below, it is visually similar to the Starsys/Orbcomm/Elippso architecture shown above.

But! More than just an enhancement to cellular service, Iridium provides its own long distance transmission network. Note that the signal from your car, truck, or bicycle telephone doesn't go through an existing long distance phone network. If, while bicycling to work at the shimmering ISSO twin towers with the commanding view from the Blue Ridge to the Potomac, you decide to phone the ISSO overland express plying the back roads of Piedmont, your call is routed up to the nearest cluster member, then relayed (cross-linked) from satellite to satellite until the call reaches the cluster member passing over Piedmont. Then it goes down either directly to the truck, or to the local phone service.

About a week after announcing the proposed Iridium system, The Wall Street Journal did a special interest story on national telecommunications systems, noting them as important export products for many smaller countries. Well, Iridium is stirring up that captive market. When you threaten steady revenue streams, you're likely to get noticed.

More complex clusters like Iridium relay communications allow cluster members to link two users anywhere on earth without terrestial links.

The last chapter of "Saga of the Satellite Clusters" has definitely not been written. We're now in the sort-of-thinking-about-putting-paper-into-typewriter stage now. Nobody knows which systems will succeed or how the world will regulate, tax, and utilize them. But the prospect of punching a button on your wristwatch,

203

Micro Space Craft

which is spending the afternoon on the beach with you in Shirahama, Japan, and dialing up your boss, who is, as it turns out, rowing his sloop into East Greenwich Harbor, Rhode Island, for repairs, is going to become a reality in the 1990s—thanks to a lot of low earth orbiting small satellites and the organizations that fund, build, launch, operate, and license them.

Chapter 17
Where to Look for Historical Underpinnings, Term Definitions, and Revolutionary Zeal Turned Up to 11

First, Some Definitions of Small Satellites

ISSO, the International Small Satellite Organization newsletter, which was where this book began, has never successfully resolved two key issues:

(1) How to pronounce ISSO, the leading contenders being as in "isometric" vs. the opposite of "is not."
(2) The origin of ISSO's most unusual contributing author, Dr. Zlitz.

To fathom question #1, and quite accidentally, to resolve, #2, the first international meeting on how to pronounce ISSO was organized cooperatively in Germany by Zeringen Landkreise, Interbopfingen (Chamber of Commerce), Tuttlingen (Gymnasium #2), and

Zugspitze (the preservation society thereof), known collectively as Gruppe ZLITZ. Subsequently the conference was renamed to Zuerst Lebendige Issoaussagensinnconferenz Tatasachlich ohne Zweifel, which was later popularized as ZLITZ. No conclusion was reached on ISSO vs. ISSO, but the opportunity was exploited to list a few popular definitions of small satellites. Meanwhile, the ISSO newsletter has been renamed *New Space*, rendering the issue moot.

From ISSO: That which the satellite establishment, including the US Government's FCC, and WARC have to date not taken into account in their planning. Those that are out of the realm of the previous legislation, regulation, and policy formulation of almost everyone. That's a small satellite.

From Herr Dkr. Professor (prepare for a lecture at the drop of a hat) R. Zlitz: "Small—that's not the question. It's cost. If the satellite is cheap enough, and you can launch it cheaply enough (hence maybe it needs to be small), then you have a lot of advantages. You can take risk. Counter example: the Space Shuttle offloaded its last ferrous memory beads in 1991, when 4 Mbit chips were already in lots of salespersons' sample kits and 1Mbit memories were at Radio Shack. The Shuttle is not cheap, cannot take risk. Small satellites are where new technologies are welcomed into space.

"Redundancy. Maybe you don't need it. A small satellite failure won't sink the treasury of a mid-sized country (refer to Risk), but more importantly, reliability is intrinsically high in simple systems. How often does your Sony Walkman, suddenly and without notice, just croak? Probably never, not counting the time you dropped it out of your beach bag into the surf at St. Kitts. When was the last time your car, even an old car, just suddenly stopped running in the middle of the freeway, not counting running out of gas? Probably not since you traded in that '65 Ford with the tube AM radio and the Bug Eye Sprite made that final journey to Lucas in the Sky with all of your Diamonds. The point is, a system with relatively few components suffers relatively few failures. If you want redundancy, fly two.

Chapter 17—Historical & Cetera

"Small satellites are cheap, right? That's it. If it's cheap enough that a failure isn't fatal—and you can use modern components to keep parts count low so that you don't mind single string (non-redundant) systems—it's small, whether it means a satellite, launch vehicle, or brewery. Speaking of which, there's a microbrewery just down the street…"

From DARPA (aka ARPA, Land of Concise): Small means smaller than what people were using before to do about the same job. Doesn't matter if it's 100 kg or 1000 kg.

From a Military User: Single mission. Big satellites carry multiple payloads. The mass of the interface control documents exceeds that of the spacecraft and approaches the weight of the launch vehicle. Little satellites mean you can afford a lot of them, and systems are composed of a network of small satellites instead of one or two big ones. That means hardness, so that destruction or failure of one may mean degradation of the system, but not total loss of capability. Small satellites means multiple satellites doing missions previously done by smaller numbers of larger ones.

From Scientific Users: Something you develop in two years instead of 10. System simplicity means fast development time, which means you get research done while someone other than your grandchildren still cares about it.

It also means the science payload is in charge, not a guest. Scientists dictate where the spacecraft is pointing, when the payload is operating, and how to optimize the use of on-board resources like power, computation, and memory space.

SMALL. We're talking small.

From the Program Manager: It means the program organization is small enough to avoid documentation beyond the actual design drawings—schematics, layouts, and block diagrams the engineers need to build their systems. Systems engineering, quality control, test, integration are all done by the engineers building the system. They all understand the whole system, and cross-disciplinary engineering is the rule, not the exception. Built by a team where everyone, or no one, is a systems engineer, there are few design rules. Each engineer selects materials, components, design, and fabrication methods that meet the top level program objectives. If I can manage like that, it's a small satellite.

From the Client/Customer: It means keeping the faith. Small satellites put a lot of emphasis on the individuals building them, instead of on the bureaucracy that controls them. You need to use a strong, capable, diverse team and let them perform with a minimum of bureaucracy. Burdening the program with the reporting I'd otherwise like enlarges the team and the cost. Pretty soon, the development group is too big to communicate intimately, you need paper interfaces, you need people to read and write specs, people to go to meetings, people to enforce quality and standards, and money to pay them. And you don't have a small satellite program anymore. All in all, not the most comfortable thing to purchase—you can't hide behind the way it's always been done. The small satellite is a way to get into space where the alternative, on my budget, is not to fly at all.

From the Engineer: The design process is interactive. You've got a lot of constraints. You can't provide hundreds of watts of power, microradian pointing over 4 pi stearradians, cooling to the helium lambda point, a gigabit per second downlink, a terrabyte of memory , and a couple of Cray-II equivalents to handle it all. When the reality finally dawns on the customer, a conversation starts that lasts almost throughout the program. It can get unpleasant for everybody as we grapple with the various "boxes"—mass, power, volume, cost—whose constrains we have to live within. The user sometimes has to downscope expectations on the number of payloads that can be used, how big they

Chapter 17—Historical & Cetera

can be, and also on the quality of the components and the redundancy. It's difficult and, to keep costs low, we can't spend forever on every decision. In the end maybe nobody gets exactly when they wanted at the subsystems level, but everyone gets an efficient, low cost system — 80% of what they wanted for 20% of what they would have paid, and flown in 20% of the time needed for a conventional satellite program. It's a compromise users are making more and more.

Small Spacecraft Time Capsule: The Way (We Think) We Were

Meandering among the hundreds of attendants and rooms full of industry displays both from small organizations and almost all the major aerospace companies, the sales representative men in their ties and jackets and women in dresses and makeup all crowded into the Utah small satellite conference mostly selling whatever could be sold to a sea of other sellers, Sven Grahn from Swedish Space Corporation lamented to me: "I don't think this bunch needs us any more."

Achieving a state of total redundancy might be the only sensible goal of adulthood. In the MTV world of news bytes and bungee jumping, it may be the only justification, outside of a challenge to health care technology and economics, for the existence of us over 30-somethings.

Around the time our so-called industry congealed out of the primordial soup, Utah was a bunch of engineers, most of them from Amsat, giving talks on satellites we had built or hoped to build depending on how much money we could steal out of the cookie jar at home or how many parts could be salvaged from the stores warehouse back at school. The industry displays were some Amsat models on card tables. There was no industry. That was 1986. The Air Force Chief of Staff had yet to come out with the simplifying declaration of the decade—that there was absolutely no use for these small satellites and that they were a waste of time and resources.

Nonetheless, a few military men and women came to Utah in civilian clothes, just to play it safe.

Just to remind you what 1986 means — that era was also before laptop computers; many companies didn't yet have fax machines (considering them, one presumes, a waste of resources too); and it was also before Reagan's military buildup reversed the first derivative, making over 100,000 aerospace engineering and management jobs redundant.

The notion that a small satellite is just a small large satellite is a sorry devolution of the initiative we all undertook to found this field. Think how far we've all come from Victorian England and the Puritans, where for the sake of some abstract ideology men and women bound their bodies in corsets and tight boots, in long jackets and puffed out dresses, and in some pretty monstrous neckware and heavy duty hats and plumage. While the monotonic trend over the past 250 years has been toward informality, Utah has gone from the land of Jeans to the land of Suits.

So Sven, I don't think we're totally redundant yet. We have yet to leave behind a few strands of small satellite DNA encased in amber that perhaps will enable some mad scientist of the 21st century to rebuild what the crush of commerce, busily saving jobs at the aerospace dinosaurs, seems so intent on ignoring. Hence and forthwith, some organic remnants of a great endeavor, building cool little satellites, often without money.

KISS. What does it really stand for? Advice given to English Barristers in training: Keep It Simple and Salient. Nobody says salient anymore. Most people think it has to do with an ionic solvent widely used by chemists, biologists, and contact lens wearers, a sorry fact when you then must experience the products of the generation of lawyers that has forgotten it.

Radcal, ALEXIS, Freja, Healthsat are all up there working away, 24 hours a day, each built with about the budgetary equivalent of a conventional satellite subsystems design review meeting.

Chapter 17—Historical & Cetera

The almost perfect success record of small satellites is not a result of our supreme talents, unfortunately, but rather of the application of KISS. When you build a satellite that weighs 50 kg, you have about 1% the parts count of a satellite weighing five tons. Ignoring redundancy considerations (and both small and large satellites have redundant systems, making the argument itself somewhat redundant), the probability of a critical component failure in a conventional satellite is, assuming we're using 99.99% reliable parts, about 1000 times greater than in a small satellite. Experience bears this out in everyday life. What's the probability of your car collapsing on the way in to work this morning? Not too great, assuming you're not driving a '69 MGC equipped with full Lucas electrics. But what's the probability of seeing two or three cars stranded along the roadside during your 10-mile commute? Virtually certain. For large numbers of cars, representing in total a huge number of parts, several are always going wrong, but for small numbers, their reliability is remarkable, especially remarkable for us graduates of the Ford 289 V-8 school of 1960s driving technology.

The wonderful reliability of portable radios, and for that matter of laptop PCs, is again a result of their very low parts count. My walkman AM/FM has exactly one IC chip, two connectors (battery and headset) and three knobs - on/off, volume, and tuning. It refuses to quit, but my Drake TR-4cw ham radio transceiver, with over 25 vacuum tubes and multiple circuit boards loaded with discrete resisters, capacitors, and inductors, a panel full of knobs to tweak and over 100 pins of connectors on the back, breaks about bimonthly. If I don't use it too much.

We small satellite mammals have just picked a much, much easier problem. Our reward has been excellent reliability, perhaps despite our lack of extreme attention to quality control and detail. Biggest hazard? Getting underfoot of big mammals!

LSMFT. Does it really just stand for Lucky Strike Means Fine Tobacco? Or is it Lighter Spacecraft Mean Faster Technology Transfer? Probably the former, but when a client drops a significant fraction of the GNP on your RFP, you can bet the war cry goes up:

Micro Space Craft

Space Qualified! What does space qualified mean? Like the rest of society, the space world lives in its euphemisms, much as the Sufis live within their own contradictions rather than face those of a world of hunger and disease. In fact, space qualification is pretty easy. You test something through the range of environments it will see in space. Computer chips and software really don't care much about zero gravity, so putting them through thermal vacuum, vibration, and maybe radiation testing pretty much constitutes space qualification.

The cult of space qualification prefers that every damned part on the satellite has flown in space before. Clearly in 1957 this was tough. The only things "space qualified" according to this definition were comets, asteroids, planets, and their moons, stars, and some dark interstellar media, none of which are all that useful for telephoning your aunt in Poughkeepsie from vacation in Pago Pago. At some point, roughly when Apollo 13 limped back from the moon, NASA collected the toys of its youth and decided those would be the only toys of its adulthood. Imagine living on a diet of Mousetrap and Swinger for 60 more years? NASA's world has no Trivial Pursuits, no Bungee jumping, no Where in Time is Carmen San Diego? Just Monopoly and Scrabble.

Why else would the Space Shuttle computer system have been dwarfed by the awesome computing gazorch of an old HP-35 calculator carried into space by one of its crew? Is hiring a team of quilt makers from West Virginia to weave together ferrous beads a better way to build computer memory than integrated circuitry? The Space Shuttle made do with 2000 bytes of RAM, about 0.1% of what's built into the computer your kid uses to play Kung Fu Cat on. But that 2k was Space Qualified!

The problem with space age retrotechnology is that it drives parts count up enormously. Failure probability goes as the product of the reliability of the critical parts. A single part might be 99.99% reliable, or 0.9999 in probability theory parlance. Two such parts in series have 0.9999 x 0.9999 = 0.9998 reliability, or 99.98%. One hundred such parts have reliability 99%, and one thousand of such parts in theory have reliability 90%, a significant 10% failure risk.

Chapter 17—Historical & Cetera

The figure below illustrates the overall reliability of a spacecraft vs. its part count for three classes of individual component reliability. The miracle isn't that little satellites are so reliable, but rather that big ones work at all.

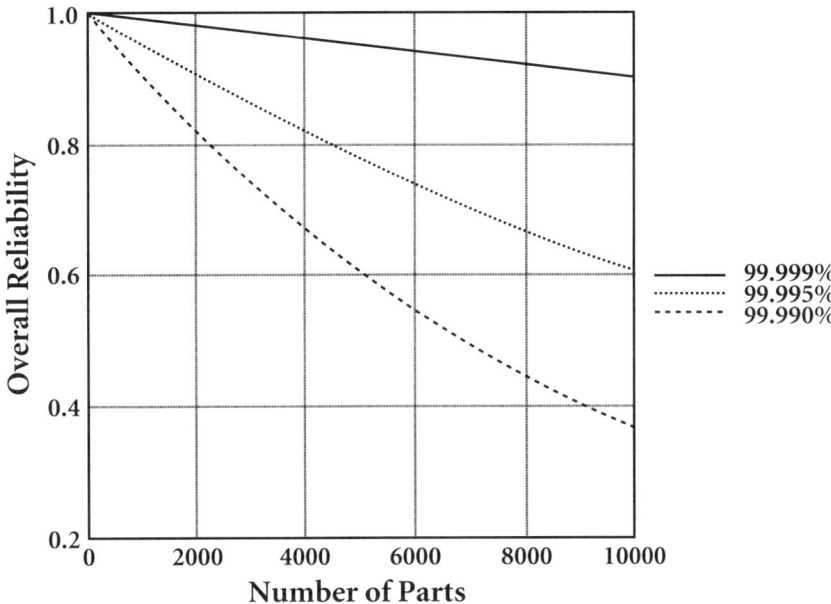

A simple system with a few parts is more reliable than systems with more parts — even if the simple system is made with inexpensive, less reliable parts.

The "What's the worst thing that can happen" reliability corollary. When the space shuttle went down, the loss was measured in human lives, discussed and debated in our national news, scrutinized by Congressional inquiry. A serious blow was dealt to the US program in space, indeed to our national prestige. The Shuttle, then our only means of launching a large fraction of our space missions, was grounded for two years. This was the equivalent for space of the gas lines of the mid '70s. Like those gas lines, it has faded from memory for most of us. But then when Mars Observer went reclusive on us, the loss was measured in billions of dollars and thousands of engineers' and scientists' careers that had been dedicated to the Mars mission. Once again society questioned our leadership in space. The interesting reaction from at least some quarters at NASA was "no more major missions." Really? Then why do we need 20,000 people under one roof at NASA?

By contrast, the failure of a $1M minisatellite is a small paint scratch— 0.08% of NASA's budget, and maybe 0.03% of our total national space budget. (Yes, Virginia, the CIA and NSA have space budgets too, as do NOAA, DoD and DoE. NASA is no longer THE space agency; it is A space agency.)

The initiative by USRA to launch two space science missions for $24M is flawed in that it's too small. USRA should press NASA for $240M to do 20 missions—each year. They are built inexpensively, largely by students. Hundreds of graduate and undergraduate students design, build, test, fly, and operate satellites within the span of their scholastic careers. They undertake significant technical risks to do innovative science in new ways. They won't be tested to extremes and their designs extensively reviewed. Even if they achieve failure rates unprecedentedly high in small satellites, like 10% or 20%, the nation will succeed in launching 24 successful science missions a year on 2% of NASA's budget.

The cruel history of life on earth is that a lot of us don't make it. Mortality is the rule. Yet we survive because of the enormous number of attempts we make at life. An insect might lay 10,000 eggs, and an individual eco-niche harbors millions of similar insects. Not every one can succeed, but enough do succeed to make insects one of the most successful species on earth, both in terms of longevity and sheer numbers. The way to do small satellites is not one at a time. It is 5 or 10 or 100 at a time. Accept failure rates of 10% or more. By flying 11 satellites to get 10 in operation on orbit, you can bypass most of what makes satellites so expensive, such as high quality parts, redundant systems, extensive testing of each unit, restrictions to use only "space-qualified" components. With these steps, satellites can be built for 1/2 or less the cost, and a 10% reliability tax is lost in the noise.

Chapter 17—Historical & Cetera

What's the difference between a drunk and an alcoholic? Drunks don't have to go to all those classes. But one thing they teach at those remedial classes is that you can't do everything. Small satellites don't do everything. They do a few things pretty well. The way to design a small mission is not to decide to cut the mass of Space Telescope from 10 tons to 100 kg. To do so might require an investment equal to or greater than the US GNP. Low cost missions result, in part, from limiting our sights to those missions that nicely fit. The key is, fit into what? I'd say, fit into a work package for the entire mission that 15 people or fewer can accomplish. Limiting management complexity and keeping overhead low and efficiency high, the resource requirements and number of active components definitely remain modest enough to allow innovation without undue risk.

Is high technology the road to low cost? As a swimmer, I know a lot of other swimmers. Many of my swim buddies train for swimming by lifting weights, pulling on elastic cords, training on complex machines. None of these people swim faster than I do, but they are much better at lifting weights, pulling on elastic cords, and operating heavy machinery. Nothing wrong with that, but if your goal is swimming, there's nothing like practice in the pool, which is itself an interesting lesson for people who might want their own low cost satellites. You might try contacting some of the people who build them, as opposed to the people who instead build very high cost satellites. Compelling logic, don't you think?

If your spacecraft development goals are simplicity, reliability, and low cost, there is no reason not to embark on a technology development program. It won't help you achieve simplicity, reliability or low cost, but it's fun and could even pay off down the road. For instance, you could come up with a new clamp for attaching chain link fences to posts, sell it to the Department of Highways, and retire with millions. Oops, too late. Somebody already did that. Besides lots of fences still attached to their posts, this innovation allowed its inventor to open a very nice golf course too. But nowadays neither he, nor a satellite technology R&D program, turns out a cheap satellite at the end of the day, month or year, even with multi-year follow ons.

As far as building low cost satellites is concerned, we partly achieve low cost by feeding off the R&D products of huge programs that can afford to do technology development. For example, the digital hardware systems of the most advanced small satellites exploit microprocessor, memory, and logic technologies pioneered for laptop computers, a business about five thousand times larger than the low cost satellite market. Even a very generously funded technology program cannot compete with the capital resources Boeing invests in its advanced aircraft structures, or IBM and Apple in the microcircuitry of the Power PC, or Toshiba and Union Carbide in advanced rechargeable batteries. The surface mount technology built into your cellular phone, walkman, and car CD player is just now migrating into small satellites without benefit of any small satellite R&D support. We let Motorola & Co. pick up the check for that lunch.

Herewith, the advice of the ages (and aged): Go forth without shame, learn about the hot new technologies that are driving new product developments in industry, and apply your space expertise to choosing those that really offer promising improvements to your spacecraft design. High density static and pseudo static RAM have been a boon to microsatellites, and memories of 1 GBit capacity have already flown on sub-50 kg busses. Miniature hard disks, so important to the laptop market, have not migrated to space in large number. They need air tight envelopes, and they demand relatively high power. In addition, their sensitivity to the launch vibration environment (no, surviving a fall off a desk to the linoleum floor isn't good enough) and the disturbance torques their motors generate, coupled with the fact that the cost of RAM is not a driver for even the lowest cost satellite, just haven't made them a logical choice. Flash RAM, which requires no idle power and is intrinsically reliable, is now being increasingly used in lap and palm tops, as well as in PDAs, Personal Digital Assistants like the Apple Newton. Other promising technologies are smaller, lighter Nickel Hydrogen batteries, already widely used in laptops. Ultimately, rechargeable Lithium batteries will be adapted to space.

Another area ripe for scavenging is test facilities. The world is chock full of shock and vibration labs, thermal vacuum chambers and RF

Chapter 17—Historical & Cetera

ranges searching for a new *raison d'etre*. The lease/buy decision is pretty trivial here.

Fun. In the space world of the '60s, there was fun. The old guys don't like to admit it, but when they launched a rocket out of Cape Canaveral daily, they were loving it. Who wouldn't enjoy calculating, on a slide rule in real time, the trajectory to the moon for three astronauts? What would you have paid NASA to let you work in mission control during a moon landing?

Let's face it, the space world of the '70s, '80s, and '90s is no fun. We live in fear of failure. None of the good we do can offset the negative feelings loss of a mission creates. NASA is a huge bureaucracy, not a playground for the best minds in the US. Cruel, but true. What was fun in the '70s and '80s? Silicon Valley was fun. Apple was a renegade and thousands of little companies sprang up across the landscape like the first growth after a long drought. Thanks to the digital era, most of these Silicon Valley upstarts weren't even in Silicon Valley, except maybe in spirit. They are just as easy to find in Canberra or Columbus as in Cupertino.

In the '90s, Silicon Valley has gone corporate. Bill Gates' apparent megalomania ultimately seeks to destroy the competitive environment that has so strongly benefitted computer users. Apple, no longer the oddball utopia it was, offers some small refuge from lock step compatibility with the C:\ crowd. Software development is an industry now, bureaucratized and overcrowded with competitors. It's managed for productivity, not innovation, and the breakthrough applications that rocketed new companies to prominence are very few and far between. Big success stories like Quicken are more marketing marvels than new innovations in using synthetic thought.

Microsatellites are fun, at least for now. Maybe not as much fun as Silicon Valley was, but we aren't serving a consumer mass market of one billion people yet. Building our industry on a handful of customers limits applications and innovation. The thrill of building a whole satellite in a group small enough to share two pizzas and a watermelon for lunch is real and infectious. Just as Apple, AST, and other startups enjoyed conquering the computer giants who made the Cybers and the 360s, there is a competitive excitement to a company of three people walloping TRW, Boeing, and Ball in a competitive spacecraft procurement. Maybe the days of the giants aren't numbered, but their eco-niche is a little narrower.

What good is fun in the post-MacNamara era? Pragmatically, the best technical, managerial, and marketing minds gravitate to fun even more than to money. Rocketry made great leaps when our society's greatest engineers aspired to work in it, and when the toughest curriculum at MIT was Aerospace Engineering. In the last twenty years the attraction has been Wall Street and computers, at least one of which has revolutionized almost every aspect of our lives. But in a capitalist society, miracles take money. Progress resulted from a combination of sophistication in the financial and technical realms, and Wall Street sophistication is what financed the revolution in computers.

Today's microspace entrepreneurs are seeking a similar unity among financial, technical, and market forces. For remote sensing and communications, a new regulatory environment is also needed to effect a new order in services from space, as evident in the flurry of FCC filings, recent WARC proceedings, and new initiatives in commercial remote sensing. This new activity is the direct result of mixing a new ingredient into the morass of the old aerospace doldrums, a powerful boredom solvent known as…

Chapter 18
Space History and a Possible Future

According to *The Hitchhiker's Guide to the Galaxy*, the history of human civilization can be condensed to 3 lines:

- How do we survive?
- What can we eat?
- Where should we go for lunch?

Had Douglas Adams predated Will and Ariel Durant, thousands of pages and two lifetimes could have been spared. One could say the same of millions of lives and hundreds of wars. Progress marches over all of us.

It's currently marching over the world's aerospace industries and institutions. Let's review five decades in five lines:

'50s: How do we get to space?

'60s: What can we do in space?

'70s: How many bigger and more expensive things can we do in space?

'80s: Who's gonna do what in space?

'90s: How can we spend less in space?

To wit:

In the mid '50s, President Eisenhower was advised that it was impossible to put a satellite in orbit, and also worthless even if it were possible.

The '60s saw remote sensing, LEO and GEO communications, unmanned space science, and manned space flight all achieve significant accomplishments.

NASA's Space Station, Space Shuttle, Spacehab, and Mission to Mars (the latter now awaiting resuscitation at the right political moment) all came to significance during the '70s.

The great international commercial space race was the significant feature of the '80s. New countries and new competition in semi-private space monopolies commenced space activities, with Ariane, Long March, Proton, and Delta vying to launch satellites built in the US, Japan, and Europe. An ominous cloud of commercial competition appeared on the horizon of Intelsat and Inmarsat.

The '90s arrived like a New Year's Day hangover. Defense is no longer as huge an engine of demand. Nobody wants to consider a multi-trillion (with a T) dollar Mission to Mars, sticker shock and budget buster being this decade's favored Washington clichés. While many look back to our Glory Days, and try to relive them with great missions, the real challenge facing our community is three letters: ROI—return on investment. Trillions have been spent on space. Perhaps it is not too cruel to ask our long-haired teenager, loafing around the living room and watching MTV, to contribute more and to ask for less.

In ROI terms, small programs have a significant edge. ROI is benefit divided by cost. If cost goes down by factor 20, while benefit goes down by 2 or 4 or 10, you are a winner. It's doubtful anybody will soon make the magic crystals in space that NASA hoped would be worth $10,000 or $100,000 a gram. Even the shuttle could break even at those prices, but some very modest, sensible, and valuable programs now dot the small space landscape.

Chapter 18—Space History and the Future

Little LEOs. Yes. In theory, three massive GEO comsats with spot beams could provide quasi-global cellular phone performance by means of small handheld and wristwatch terminals. Fact is, that's not happening (refer to big LEOs below). Small LEO satellites providing electronic mail and messaging are already in operation serving relief workers in the Third World (HealthSat and VITA); taking environmental sensor readings (Orbcomm); and linking electronic mail as a satellite extension to Internet (AMSAT). On a commercial basis, non-real-time communication from small LEO satellites is a fact of the '90s.

Big LEOs. Iridium. Globalstar. Ellipsat. These programs and their spacecraft are so large and growing that they stretch any contemporary definition of small satellites, launch vehicles, and space activity. But their roots trace to the thesis that ROI can be increased by decreased physical size, which lowers manufacturing and launch costs. Simultaneously, program risk can be decreased by spreading performance over many independent small spacecraft.

I dismiss as naive Motorola's claims that Iridium has nothing to do with small satellites. Rather, it is a sign that an innovation has really matured when the innovation itself is taken for granted. Does Gap pay tribute to Levi Strauss? Today's rock and rollers admit no relationship to Moog. Do Apple computer users know or care about Eniac, Turing, or even Wozniak? Innovators build a kitchen, set the table, and stock the refrigerator, so that they can enjoy watching their children eat them out of house and home!

Space Science. ALEXIS, now operating on orbit; HETE, launching in 1996; Terriers, launching in 1997; and DOE and NASA's astrophysics satellites, all built by AeroAstro and all weighing under 150kg, have capabilities comparable to much larger, costlier, and more complex satellites of the '70s and '80s. Freja, Odin, and Viking, built by Swedish Space Corporation, are proving that small platforms can do leading edge astrophysics and geophysics research. NASA's Discovery program, aimed at very small missions, was inundated with enough proposals in planetary, earth, and astrophysics research to carry out an ambitious program lasting for decades or even centuries.

Remote Sensing. OSC's Seastar is a well known small satellite remote sensing program. However, at least 5 other remote sensing programs worldwide plan to use one or more small satellites with imaging and non-imaging sensors to manage crops, detect pollution, and aid meteorology. One program not at all secret is Worldview, which is placing mini-Spot class imaging devices aboard small satellites at LEO.

Education. University of Surrey, Boston University, Northern Utah State, Technion, Stanford, Colorado University, Weber State, Korean Institute of Technology, and the University of Tübingen all offer academic programs including hands-on design and/or development work for small satellites.

Ten years ago, small satellites were not even a curiosity outside of AmSAT. Nobody even thought about them. They weren't significant enough to have an opinion about.

According to ISSO's archives, there were 16 small (where small equals less than 300kg) satellite launches in 1992, 31 in 1993, and 26 in 1994. We are giving our own community, and more importantly, the community of users in science, communications, education, and remote sensing, a new set of options in addition to the large, highly capable, but costly conventional satellites. Reaching out to this much larger community of lower cost users is not so appetizing to some organizations used to munching on large budgets and long programs. But to survive as an industry may mean that survival passes to those who live by smaller mouthfuls. Having fun along the way and really moving our technologies forward to better serve our users, small satellites are providing our next step into the future.

Anybody for lunch?

Index

A

Absorbers 116
Absorbtivity 188
Absorption 121
Abstract 210
Abstractions 72
Acceleration 12-14, 36-37, 195
Access 72, 105, 143-144, 166, 187
Accident 199
Acid(s) 22, 39, 175-176, 179-180
Actuation 163, 172
Actuator(s) 41, 133, 163, 171, 192
Adaptation 146
Addition 127, 141, 181, 216, 222
Adhesion 131, 191
Adhesive 41
Advantage(s) 29, 34, 38, 42, 50-51, 73, 87, 136, 144-145, 151, 153, 163, 199, 206
AeroAstro 79, 142, 151, 153, 221
Aerobics 110
Aerosols 188
Aerospace 5-6, 46, 50, 166, 171, 178, 188, 209-210, 218
AIAA 42, 81
AIAA/USU 81
Aircraft 3, 8, 10, 35, 96, 107, 172, 216
Aircraft-inventing 100

Airlock 191
Airplane(s) 10, 20, 49, 66, 83, 124-125, 172, 175
Airplane-based 126
Airport(s) 49, 66
Alarm 58
Alchemy 67-68
Alcohol 189, 191
Alexis 77, 100, 135-136, 151, 210, 221
Alexis/MOXE/Euvita 151
Algorithm(s) 147, 151
Alignment(s) 76, 86, 157
Alkaline 173
Allen 51, 70
Allocation(s) 5, 97, 104, 201
Altimeter 73
Altitude(s) 48-51, 53, 55, 57-58, 63-64, 71, 105, 126
Aluminum 28, 32, 67, 111, 119, 158
AM 58, 83, 87-88, 91-92, 96, 101, 107, 123-124, 200, 206, 211
AM/FM 101, 211
Amateur(s) 33, 43, 164
American(s) 85, 87, 197
AMF 135
Ammonia 30, 32
Amp-hour 178
Amperes 177-178
Amplification 97, 100

223

Micro Space Craft

Amplifiers 101
Amplitude 103, 106
Amsat 43, 119, 127, 164, 209, 221-222
Analog 74, 83-84, 104
Analysis 133, 148
Analyst 46
Anatomy 19-20, 144
Anechoic 132, 165
Anode 177
Anomaly 81
Antarctica 72
Antenna(e) (s) 51, 54, 56-57, 64, 87, 92, 95-96, 99-100, 105, 107, 123-124, 127, 129, 132, 135-6, 156-160, 163-164, 166, 183, 190, 199-200
Antiprotons 80
Antistatic 194-195
Aperture 85
Apogee(s) 62-64, 135
Apollo 11, 56, 111, 172, 212
Apple 216-217, 221
Archimedesland 2
Arcjet(s) 41
Arcs 41
Arctic 164
Ariane 29, 37, 41, 220
ARPA 100, 207
Array(s) 144, 147-148, 150, 182-185
Asia 87
Asteroid(s) 53, 212
Astromast 164
Astronaut(s) 13, 37, 111, 217
Astronomers 2, 101
Astronomy 4, 96, 123
Astrophysicists 48
Astrophysics 221
Asynchronous 149
Atlantic 63, 80-81, 85

Atmosphere 11, 21, 36, 48-49, 51-53, 106, 161, 165
Atmospheric 13, 21
Atom(s) 25, 27-28, 30-31, 39, 68, 70, 80, 114, 146
Atomic 25, 68, 80-81, 86
Attenuation 106
Attitude(s) 22, 40, 43, 70-71, 75-79, 81, 111, 119, 123-124, 129, 135, 137, 143, 169, 174, 179
Automobile(s) 7, 11, 13, 20, 162, 177
Avionics 149
Axis 74-75, 127, 133-135, 137, 183

B

Balloon(s) 166, 195
Band(s) 85, 88, 92, 94-96, 140, 169-171
Bandwidth 102, 104, 108
Battery (batteries) 1, 9, 59, 66, 111, 119, 140, 161, 165, 168, 173-182, 196, 202, 211, 216
Beam(s) 56, 100, 105-107, 199, 201, 221
Bermuda 63
Bi-phase 103
Bias 130, 146
Binary 48
Biology 69
Biorhythms 17
Bipropellant(s) 30, 32, 43
Blanca 198
Blimps 133
Board(s) 10, 12, 15, 34, 40, 43, 50, 78-79, 92, 104, 107, 119, 131, 140, 144, 153, 155, 157, 191-192, 194, 211
Boeing 216, 218
Bolts 20, 119, 163, 167, 170-171

Index

Bomb 114
Bonding 27, 36, 38, 131
Bonds 11, 25, 27-28
Boosters 37
Boston 8, 16, 99, 222
Boundaries 16, 202
BPSK 103
Braun 48
Braza 37
British 32
Broadcast(s) 54, 56, 88, 95, 100
BTW 69
Buchwald 97
Budget(s) 75, 77, 83, 110, 160-162, 177, 208, 214, 220, 222
Buick 11, 157-159, 161, 166
Builder(s) 5, 73, 75, 77, 175, 201
Building 2, 5, 8, 11-12, 51, 67, 74, 160, 172, 189, 197, 208, 210, 216-217
Builds 151
Bungee jumping 209, 212
Bureaucracy 208, 217
Burma 124

C

C-cell 179
Cables 107, 177
Cabling 149
Cache 139
Cadmium 173-175, 180
Calculation 148
Calculator(s) 15, 51, 212
Calculus 4
California 40, 72
Cambridge 66
Camera(s) 51, 123, 166, 176, 196
Canada 72, 171, 202

Canaveral 196, 217
Canberra 217
Capability (capabilities) 40, 148, 161, 207, 221
Capacitors 155, 211
Capacity 12, 95, 106, 141, 143, 146, 151-153, 176-177, 216
Capsule 209
Carbon 28, 30
Carcinogen 30
Carmen San Diego 212
Cassette 86, 143
Categorization 22
Cathode 177
Causes 119, 127, 129, 147, 165-166, 180, 182-183, 196
Caution 196
Caveat(s) 79, 192
Cdma 105
Cell(s) 84, 93, 111, 155, 174-175, 177-183, 185, 188, 198, 202
Cellular 15, 84, 92, 94, 96, 145, 168, 178, 197-198, 202-203, 216, 221
Cellular phone 15, 84, 168, 198, 202, 216, 221
Celsius 73, 113
Centrifuges 132
CEOS 133
Ceramic 107
Cerillos 198
Chamber(s) 20, 24, 33, 41, 132, 158, 165, 188, 205, 216
Channel(s) 68, 95, 102, 104, 106
Charge(s) 17, 80-81, 114, 119, 140, 146, 163, 171, 179-181, 193, 195, 207
Charger 180
Charts 3
Chassis 194
Check 37, 103, 145, 147, 151, 178, 216

225

Chemical energy 4
Chemicals 22, 28-29, 31, 36, 39, 41-42, 51, 94
Chemistry 25, 67
Chemists 210
Chinese 8, 31, 33, 37, 67
Chip(s) 13, 20, 53, 146, 148, 151, 153, 206, 211-212
Chromium 67
Circuit(s) 67, 80-81, 88, 92, 105, 119, 146-147, 153, 155, 183, 191-192, 211
Circuitry 77, 147, 163, 212
Circulation 158, 192
Clean 5, 30-31, 170, 187-192, 195, 198
Cleanliness 192-193, 195
Cleanroom 181
Clock(s) 58, 83, 104, 181
Coating(s) 67, 116, 188, 191-192
Code(s) 68, 85, 103, 105, 148
Coding 106
Coil(s) 74, 76-80, 86, 132, 136
Collector(s) 120, 156
Collectors 120
Colorado University 222
Columbus 3, 73, 75, 217
Combustion 7, 25, 41, 110
Commercial 43, 146, 150, 153, 166, 180-182, 194, 218, 220-221
Commercial-grade 180
Communication 5, 95-96, 106, 221
Communications terminals 4
Compass(es) 3, 73-76, 86
Compensation 66, 102
Competition 109-110, 220
Complication(s) 29, 57, 114, 119
Component(s) 1, 30, 32, 43, 66, 74, 81, 88, 118-119, 143-144, 146-147, 150, 152-153, 161, 182-183, 187, 192-194, 211, 213-215
Compression 151
Computer(s) 3-5, 9, 17, 43, 45, 74, 94, 133, 136, 139, 141-145, 155, 160, 165, 167, 176, 188, 210, 212, 216-218, 221
Computer chips 212
Computer memory 212
Computing 80, 133, 139, 212
COMSAT(S) 40, 221
Concentration(s) 30-32, 185
Concentrator 183
Condensation 193
Conduction 111-112
Conductor(s) 67, 111, 119, 157
Configuration(s0 49, 128, 136, 149, 182-183
Connection 99, 200
Connector(s) 41, 74, 161, 211
Constellation 57, 198
Constraints 77, 141, 175, 208
Construction 58, 168
Consulting 197
Consumer 80, 160, 175-177, 179-180, 182, 217
Container(s) 19, 28, 31-32, 111, 141, 195
Contaminants 31, 187, 191
Contamination 188, 191, 194
Contributor 20
Control(s) 5, 11, 38, 40, 43, 45, 71, 77, 79-81, 94, 96, 102, 111, 114, 123-124, 127-128, 133, 141, 147, 169, 194-195, 207-208, 211, 217
Controllability 37-38
Controller(s) 102, 133, 144, 151-153
Convection 111-112, 121
Cooling 167, 208
Correction(s) 40, 43, 101, 136, 147-148,

Index

150, 153, 162
Corrosion 193, 196
Cost(s) 2, 11-12, 29, 31-32, 36-39, 40, 42-43, 50, 53, 65, 73, 80-81, 83, 119, 140-142, 144, 146, 150-151, 153, 156, 159-160, 164-165, 175-176, 180, 182, 189, 200-201, 206, 208-209, 214-216, 220-221, 222
Counters 110
Course 3, 13, 17, 43, 67, 103, 109, 111, 114, 133, 140, 148, 171-173, 175, 196, 200, 215
Coverage 40, 54, 105, 197, 201-202
Coverings 187-188, 195
Covers 191-192, 196, 201
Craft 11
Cranston 202
Cray-II 208
Cray-XMPS 45
Crystals 88, 220
Customer 79, 81, 208

D

D-cells 178, 180
D-size 173
Data 4-5, 54, 81, 95, 99, 101, 103-104, 139, 141-154, 159, 167
Decomposition 16, 30-32
Degradation 116, 146, 151, 182-183, 185, 189, 195, 207
Density (densities) 10, 100, 126, 144, 146, 151,153, 180, 216
Deployable(s) 150, 156-160, 161, 163-164, 166, 172, 183
Deployment(s) 157, 159, 163-164, 166, 172

Desaturation 132
Design(s) 25, 39, 46, 52, 81, 102, 110, 119, 121, 123, 126, 129, 133, 144, 153-154, 156, 159, 162, 165, 167, 169, 175, 180, 182, 208, 210, 214-216, 222
Designers 11-12, 14, 81, 132, 177
Detection 101-103, 147-148
Detectors 111, 123
Development 12, 29, 38, 100, 142, 151, 153, 159-160, 162, 165, 176, 191, 196, 207-208, 215-217, 222
Device(s) 4, 7, 9, 13, 20, 36, 56, 74, 77, 87, 94, 141-146, 151, 155-156, 159-163, 167, 172, 196, 222
Devolution 210
Digitalization 83
Diodes 183
Dipoles 68, 70, 86
Dish(es) 55-56, 64, 100, 200
Disk(s) 56, 89, 136, 143-144, 167
Disposables 177, 179-180
Distance(s) 15, 51, 55-57, 86, 91, 96, 99, 101, 119, 125, 185, 203
Documentation 150, 208
DoD 214
DoE 214, 221
Doppler 97, 101-102
Douglas 175, 219
Downlink 105, 123, 141, 154, 167-168, 208
Downlinking 99, 154
Drawings—schematics 208
Dress-for-success 124
Drift 39
Drive 4, 7-8, 10, 13, 15, 46-47, 79, 86, 100, 144, 150, 198
Drive-ins 168
Driver(s) 16, 32, 74, 79, 139, 216

227

Drop 3, 16, 22, 63, 123, 143, 196, 206
Duration 31, 77, 167
Dynamic(s) 4, 81, 112, 144

E

Earth 1, 4, 10-11, 14, 21, 33, 38, 40, 46-59, 61-64, 70-72, 75, 79, 81, 83, 99, 101, 105, 108, 113-118, 120, 125, 127-129, 132, 135-136, 140, 149, 157-158, 165, 167, 174, 197, 199-202, 204, 214, 221
EDAC 147-148
Effects 13, 20, 27, 70, 92-94, 100, 111, 126, 146, 180, 183
Einstein 1, 10
Einsteinville 2
Ejection 20
Electricity 2, 67, 76, 86, 88, 106, 136, 174, 176, 193
Electrodes 175
Electromagnet(s) 76-77, 80, 86, 132
Electronics 11, 59, 80-81, 96, 111, 165, 167, 176-177, 191, 193-195
Electrons 41, 80, 86, 146, 157
Ellipsat 201-202, 221
Energy 4, 12, 17, 25, 27-28, 41, 50, 53, 70, 86, 88, 94, 99-100, 120, 140, 163, 184-185
Engine(s) 7-8, 12, 14-17, 19-22, 25, 28-29, 36-39, 41, 62, 76, 84, 87, 174, 178, 220
Engineer(s) 2, 5, 7, 12, 19, 36, 43, 68, 72-73, 83-84, 92-94, 128, 132-133, 136, 156, 159, 165, 169, 191, 200, 208-209, 213, 218
Engineering 1, 4-5, 11, 41, 46, 50, 84, 151, 171, 197, 208, 210, 218

England 95, 210
Environment(s) 5, 59, 116, 120-121, 123, 132-133, 143, 146, 150, 159-161, 165- 167, 172, 176, 180, 182, 187-1923, 196, 212, 216-218
EOL 182
EPA 31, 84
EPROM 144-145
Equation(s) 14-16, 39, 45, 117
Equator 55, 57, 63-65, 76, 129, 140
Equilibrium 72, 118, 133
Equipment 38, 42, 100, 133, 187
Equivalent(s) 5, 11, 19, 95, 99, 125-126, 197, 208, 210, 213
Error 101, 136, 147-148, 150, 153, 191
ESD 193, 195
Europe 37, 87, 220
Exhaust 15-16, 19, 28-30, 32, 34, 36
Explosive(s) 163, 167-168, 170-171
Exposure 49, 94, 135, 146, 196

F

Fabrication 188, 208
Factor(s) 11, 31, 37, 72, 88, 99, 100-101, 103, 109, 139, 157, 185, 220
Factory 35, 58
Fahrenheit 73, 113, 159
Fairchild 151
FCC 218
Feature(s) 74, 92, 95, 100, 106, 123, 128, 143, 158, 220
Feldspar 88
Ferromagnetic 79
Ferrous 30, 70, 206, 212
Field 2, 4, 13, 23, 32, 41, 67, 70-71, 73-81, 86, 125-127, 132, 160, 166,

Index

179, 210
Filter(s) 75, 101, 104, 188, 195
Fire(s) 9, 29, 84, 93, 163, 170, 174
Flashlight(s) 85, 94-95, 99, 168
Flight(s) 49, 83, 102, 119, 142, 146, 151, 156, 161-162, 171, 192, 220
Flight-proven 163
Florida 196
Flow 14, 20-21, 33, 76-77, 86, 114, 157, 163
Fluctuation(s) 73, 76, 101, 118, 121
Flux 41, 73-74, 113-117
Flying 55, 68, 89, 157, 167, 170, 179, 192, 196, 214
FM 51, 92, 96, 101-103, 199, 211
Focus 1, 4, 11-12, 97, 100, 105
Force(s) 8, 13-14, 19-20, 28, 36, 40, 73, 75-76, 79, 125-126, 130-131, 143, 157, 159, 163, 166, 169, 188, 209, 218
Form 30, 86, 96, 104, 158
Formulation 179, 206
Fraction 11-12, 14, 39, 114, 134, 192, 211, 213
Freja 210, 221
French 36, 196
Freon 32
Frequency (frequencies) 5, 86-89, 92, 94-97, 100-106, 108, 111, 148
Friction 133, 157
Frictionless 165
Fuel 11-13, 17, 20, 22, 30-36, 38, 41, 83, 198
Function(s) 68, 71, 73, 148, 151
Fusion 67-68, 80

G

G-loading 195
G-loads 159
Gagarian 47, 54
Gallium-arsenide 183, 185
Gamma 84, 91, 95
Gas(es) 8, 10-11, 13, 15, 19-22, 25, 27-28, 32-34, 36, 39, 43, 93, 128, 176, 179-180, 195, 206, 213
Gasoline 7-8, 16-17, 28, 109
Gates, Bill 74, 155
Gausstown 2
Gear(s) 7, 31, 87, 172, 189
Gearheads 7, 24
Generation 69, 106, 142, 153, 162, 187, 210
GEO 39, 54, 57, 61-63, 71, 197-198, 220-221
Geometry 4, 24, 165
Geophysics 221
Geosynchronous 38-40, 52, 54-57, 61-65, 71, 105, 116-117, 136, 140, 182, 197, 199-201
Germany 205
Glass 83, 161, 188, 196
Globe 3, 58, 96, 197-198
Goddard 48
Government 72, 89, 100, 200
GPS 3
Gradient(s) 70, 119, 125-127, 133, 135, 137, 161, 164
Gravity 1, 32, 41, 49, 53-54, 70, 119, 123, 125-127, 132-133, 137, 150, 164-165, 169, 212
Ground 12, 22, 28, 33, 38, 40-41, 47, 49, 51, 53-56, 64, 73, 80, 97, 99-106, 123, 126, 140-141, 149-150, 158, 161, 165-167, 174, 194-195, 201

Micro Space Craft

GTO 62
Gyrocompasses 3
Gyros 166
Gyroscopes 129

H

Hardware 33, 39, 107, 148, 197, 216
Hazard(s) 93, 163, 191-192, 196, 211
Health 92, 94, 131, 176, 209
Healthsat 210, 221
Heat 17, 30, 48, 67, 70, 72, 99, 111-121, 184
Heater(s) 4, 15, 86, 112, 119, 189
Heating 16, 41, 54, 94, 114, 119, 163
Helium 13, 28, 32, 166, 208
Hertz 87-88
Hete/quen 153
HF 87, 91, 95-96
History 8, 141, 157, 172, 214, 219
Hohmann 63
Holden 107
Horizon 2, 99, 220
Hubble 62
Hughes 136
Humidity 193-195
HVAC 115
Hydrazine 30-32, 43
Hydride 176
Hydrocarbons 32, 188
Hydrogen(s) 8, 15-17, 27-31, 33, 39, 96, 175-176, 179, 216
Hypergolic 29, 31
Hyperthermia 114

I

IBM 139, 216
Ice 2, 5, 42, 164, 199
ICS 182, 194
Ignition 31
Illumination 182, 184-185
Illustration 19, 55, 134
Images 4, 86, 94-95, 104, 140-141, 172
Immersion 161
Impact 2, 47, 150, 185
Impulse 13, 20, 134
Inclination 57, 64-65
Inductors 79, 211
Industry 112, 115, 143, 180, 209, 216-217, 222
Inertia 134
Inertial 133, 135-136
Infrared 86, 91, 111
Inmarsat 220
Instability 157
Installation 142
Instrumentation 3
Instruments 49, 72-74, 111, 120, 151, 160
Insulation 118, 161
Insulators 114, 157
Integrated 67, 80-81, 155, 212
Integrated circuitry 212
Integration 141, 166, 181, 196, 208
Intel 152
Intelsat 220
Interbopfingen 205
Interface(s) 144, 149, 152-153, 207-208
Interference 105-106
Interferometer 110
Internet 89, 221
Interplanetary 1, 40, 43, 72, 119, 154
Interplanetary spacecraft 1, 40, 72

Index

Interstellar 10, 128, 212
Ionosphere 95
Iridium 202, 221
Iron 30, 67-68, 70, 74, 77-78, 125
Isolation 164
Isotropic 99
ISP 13-14, 16-17, 19-20, 22, 25, 27, 29-32, 35, 37-41
ISSO 203, 205-206
Ithaco 74, 77

J

Japan 50, 196, 204, 220
JPL 45
Jungian 66
Jupiter 72

K

Kagoshima 196
Kalman 75, 81
Kanji 48
Kelvin 110, 113
Kerosene 32-33, 39
Kourou 196
Kovar 79

L

Laser(s) 106-107
Latitude 201
Launch 9, 11, 32-35, 37-41, 49, 51, 70, 81, 104, 125, 140, 143, 150, 161, 163, 166-172, 174, 176, 181, 187, 190, 192, 196, 198-199, 204, 206-207, 214, 216, 220-221
Launchers 51
Launching 5, 33, 36-37, 40, 46, 49, 53, 213-214, 221
Leak(s) 34, 93, 176, 179
Leak-proof 179
Leakage 176, 179
Leaking 34, 191
LEO(s) 55, 57, 61-62, 71, 75-77, 80, 102, 106, 108, 117-118, 140, 145-146, 167, 174, 176, 182, 199, 220-222
Light 10-11, 14, 17, 28-29, 41, 56, 63, 68, 78, 85, 89, 91, 99-101, 106-107, 113-114, 128, 174, 176, 194
Link(s) 56-57, 95, 97, 99, 101-102, 106, 115, 140, 149, 202, 202, 215
Linked 20, 27, 58
Linking 57, 107, 221
Lint 187, 189, 191
Liquid(s) 22, 27-29, 31-39, 70, 175, 179, 191
Lithium 176, 216
Load(s) 13, 36, 59, 143, 150, 160, 163, 165-166, 169, 174
Logarithm 15
Logic 27, 155, 163, 215-216
Longitude 58, 140, 201
Lubricant 159
Lubrication 157-158, 161
Lunar 2, 52-53, 200

M

Macintosh 144
Magellan 75

231

Magnet(s) 68-70, 72-73, 75, 77-79, 86, 127, 132
Magnetic 68, 70-78, 80-81, 86, 125-127, 132, 136-137, 160
Magnetism 67-71, 76, 78-79, 81
Magnetization 74, 76, 78
Magnetometer(s) 73-74, 78, 81, 156
Magnitude 20, 74, 79, 102, 150
Manacle 169-171
Manchester 107
Maneuver(s) 62-65, 143
Marman 169-172
Mars 11, 38, 213-214, 220
Mass 8, 10-14, 16, 20, 27-30, 39-41, 45, 48, 77-78, 111, 114, 126, 137, 141-142, 152-153, 165, 168, 175, 177, 179, 182, 207-208, 215, 217
Material(s) 10, 19-20, 22, 28, 30, 48, 67, 72, 79-81, 86, 94, 107, 114, 116, 118, 157-159, 160-161, 164, 166, 181-182, 185, 187-188, 191, 194-195, 208
Matter 10, 16, 51, 63, 70, 118, 130, 171, 180-181, 197-198, 200, 207, 211
Mean time between failures 2
Measure 3, 12-13, 19-20, 25, 74-75, 78, 109, 114, 133
Mechanics 5, 25, 45, 50, 52-53, 66, 197
Mechanism(s) 112, 157, 159-160, 163-166, 165-167, 170-172, 196
Memory 42, 139-155, 167, 192, 206-208, 212-213, 216
Message 68, 89, 101, 104, 200-201
Metal(s) 30, 32, 67, 81, 111, 143, 158-159, 164, 169, 175-176, 179-180, 194
Micro 110, 140
Microcircuitry 216
Microcosm 120

Microelectronics 146
Micromachines 88
Micron 156
Microprocessor(s) 20, 119, 139, 148, 151-153, 216
Microradian 208
Microsatellite 140, 154, 216-217
Microspace 218
Microswitch 172
Microtesla 73
Microwave(s) 86, 91-96
Military 31-32, 34, 37, 153, 207, 210
Miniature 88, 216
Mission(s) 11, 31, 33, 37, 40, 42-43, 45, 56, 123, 141, 148, 154-157, 160, 162, 167, 181-182, 193, 207, 213-215, 217, 220-221
MIT 45, 168
Mitchell 170
Model(s) 19, 72, 75-76, 124, 209
Modulation 3, 85, 97, 102-104, 106
Molecule(s) 8, 25, 27-32, 39, 41, 50, 92, 94, 96, 146, 188
Momentum 42, 100, 121, 129, 131-132, 134, 136, 143, 157, 161, 165-167
Monopropellant(s) 30, 41, 111
Moon 1-4, 16, 47, 52-54, 63, 175, 200, 212, 217
Motion(s) 26, 39, 47, 55, 64-65, 88, 100-101, 125-126, 129, 131-133, 157, 163, 192
Motor 19, 22, 33, 36-38, 62, 131, 157, 159, 178
Motorola 197, 202, 216
Motors 36, 41, 131-132, 142, 165, 216
Murphy 130, 157, 192

Index

N

Nanotesla 73
NASA 2, 8, 100, 168, 212, 214, 217, 220
Navigation 2-3, 9, 42, 75, 81
Network(s) 57, 85, 183, 202-203, 207
Newton 10, 216
Newtons 131
NiCad 119, 173, 175-177, 179-180, 182
Nickel 173, 175-176, 216
Nickel-hydrogen 176
NIH 175-180
Nitrogen 13, 22, 27, 30-32, 39, 43
NOAA 99, 214
Nozzles 13, 28, 38, 131, 195
Nuclear 4, 58, 68, 175, 199
Numbers 13, 17, 32, 39, 49, 73, 85, 114, 147, 177, 207, 211, 214

O

Ocean(s) 2-3, 11-12, 73, 84, 96, 128, 141, 163, 198
Odin 221
Odyssey 66
Operations 96, 188, 190
Operator(s) 85, 95, 188
Optics 4, 88, 157, 166, 188, 191-192
Option(s) 23, 38, 73, 140, 149, 160, 172-173, 175-176, 180, 191, 222
Orbcomm 164, 200-201, 221
Orbit(s) 3-5, 10-11, 22, 32-33, 38-41, 43, 45-58, 60-66, 70-71, 75-77, 99, 101-102, 105-106, 111, 116-117, 123, 126-127, 129, 132-133, 137, 140-141, 148-149, 156-160, 166-169, 174, 178, 187, 192, 197-198, 200-201, 214, 220-221
Orientation 77, 125, 128, 132-134, 165, 182, 185
Oscillations 119, 125
Oscillators 111
Outgasses 157, 164
Outgassing 185, 192
Output 74, 149, 175, 183-185
Overhead 2-3, 38, 42, 101, 202, 215
Overheating 111
Oxidizer(s) 11, 17, 20, 22, 30-32, 32-35, 38-39, 41
Oxygen 2, 29-33, 38-39, 121

P

Packaging 77, 146, 153, 195
Paint 172, 187, 191, 214
Panel(s) 59, 123, 135-136, 156, 160, 163, 165, 168, 171, 174-175, 183-184, 196, 211
Parity 147
Part(s) 1, 3, 9, 12, 15, 17-18, 20, 32, 53, 57, 59, 63, 69, 74, 101, 116, 119, 125-127, 142-143, 145-148, 150-151, 155-156, 160-165, 168-170, 180, 195, 197-198, 202, 207, 209, 212-215
Particles 13, 29, 32, 41-42, 61, 80-81, 187-188, 192
Passive reflector 1, 175
Payload(s) 11-12, 14, 36, 38, 49, 119, 124, 135-136, 160, 169-170, 207-208
Pegasus 34, 37, 49
Pentium 139
Penumbra 177

Perigee 62-64, 178
Perihelion 178
Peroxide 15-17, 31-32
Perturbations 45, 101
Phase 34, 70, 103, 106
Phone(s) 15, 17, 25, 28, 52, 54, 83-84, 92, 94, 104, 145, 168, 198, 202-203, 216, 221
Photograph 152
Photon 8, 121, 141
Photovoltaics 1, 182, 191
Physics 68-69, 81, 91, 106, 133, 170
Plane(s) 11, 54, 57-58, 64-65, 175
Platform(s) 3, 112, 136, 143, 221
Polar 57-58, 64, 99, 127
Polar-orbiting 64
Polarity 78
Pole(s) 55, 57-58, 64, 72, 74, 76-77, 81, 127-129, 196, 199
Pollution 72, 84, 180, 222
Position(s) 63, 71, 109, 141, 201
Power 4-5, 17, 22, 42, 56, 59, 74, 77-80, 83, 87, 91, 94, 96-97, 99-101, 106, 108, 113, 115, 119, 135-137, 139, 141, 143-147, 149-154, 161, 174-175, 177, 179-180, 182-185, 202, 207-208, 216
Preamplifiers 111
Precision 162
Preparation 165
Preservation 206
Pressure 19-22, 24, 121, 126, 161, 179-180, 187
Probability 148, 211-212
Procedures 188, 195-196
Process(es) 4, 7, 10, 29, 38, 67, 104, 124, 145, 148, 164, 166, 172, 182, 189, 192, 195, 208
Processor(s) 145, 148, 154

Product reliability 212
Production 5, 67
Propellant(s) 8, 11-17, 20-22, 25, 27-33, 32-42, 64-65, 76, 168
Propulsion 8, 10-11, 13-14, 22, 27-29, 33, 39-43, 125, 131-132, 170
Provisions 31
Pulsar 128
Pyrotechnics 163

Q

Qualification(s) 159, 162, 171, 181, 212
Qualified 160, 212, 214

R

R&D 83, 159-160, 215-216
R&R 202
Radar 126
Radcal 210
Radiant 112, 120
Radiate 87, 94, 112, 116, 184
Radiating 113-114
Radiation 50-51, 85-86, 91-94, 111-114, 116-118, 120-121, 144-146, 150, 152-153, 161, 165, 167, 182, 185, 212
Radiation testing 150, 212
Radiator(s) 114, 120, 165
Radio(s) 1-5, 10, 17, 43, 51, 54, 56, 68, 84-89, 91-97, 99-104, 106-107, 111, 120, 157, 165-166, 168-169, 174, 176, 179, 190, 196, 199-201, 206, 211
Radio receivers 4, 120
Radio repeater 3

Index

RAM 143-144, 148-150, 167, 212, 216
Random 86, 101, 143-144, 148
Range(s) 7, 13, 88-89, 92, 102, 108, 111, 136, 140, 150, 159, 162, 165, 176, 189, 200-202, 212, 217
Rate(s) 14, 20, 33, 55, 74, 87, 95, 101, 104, 113-114, 129, 132-133, 146, 148-153, 156, 192, 214
Ratings 140
Ratio 16, 116
Reactants 17, 29-30
Reaction(s) 10, 28, 30, 32, 41, 129-132, 156, 214
Reactor 80
Receiver(s) 3-4 85, 99-104, 120, 199
Receiver/absorbers 120
Reception 104
Recharge 168, 180
Rechargeable(s) 173, 175-177, 179, 216
Recharging 202
Recording 197
Records 104, 179
Recycling 181
Reduction 153
Redundancy 155, 182, 206, 209, 211
Reflection 95-96
Reflector 1, 126, 175
Rejection 24, 68
Relay 1, 54, 56, 94, 141, 201
Reliability 142-143, 150-151, 156-157, 162-163, 182, 206, 211-215
Remote 5, 96, 120, 191, 218, 220, 222
Remote sensing 5, 120, 218, 220, 222
Remote sensing communications 218
Requirement(s) 33, 135, 144-145, 163, 177, 182, 192, 215
Research 29, 32, 68, 94, 106, 175, 207, 221
Resistance 77, 133

Resistors 155, 211
Resonances 88
Resonators 94
Retrotechnology 212
Revolution 58, 108, 140, 218
RF 100, 111, 216
RFP 211
Robocop 47
Rocket(s) 8-17, 19-25, 27-30, 31-42, 45-46, 48-49, 51, 53, 62-63, 65, 76, 100, 104, 111, 131-132, 136, 144, 160, 168-169, 171-172, 174, 191, 200, 217
Rocketry 9, 11, 13-14, 32, 34, 36, 218
Rotation(s) 27, 71, 105, 129, 131, 133

S

Saturn 4, 11, 37, 162
Scale 12, 18, 25, 70, 180
Schonstedt 74
Science 5, 69, 99, 207, 214, 220-222
Scientist(s) 2, 8-9, 21, 26, 73, 91, 207, 210, 213
Screening 150-151, 182
Security 3, 73, 118
Seinfeld 139
Semiconductor 143-144
Sensing 5, 120, 129, 218, 220, 222
Sensor(s) 73, 75, 133, 141, 157, 221-222
Separation(s) 125, 169-171
Series 37, 39, 172, 178, 183, 212
Sextant 3
SHF 89, 91-92, 96
Shock 24, 163, 167, 195, 216, 220
Shortwave 95-96
Shuttle 13, 22, 29, 37, 50, 62-63, 126, 151, 161, 168, 206, 212-213, 220

235

Signal(s) 1, 51, 56-57, 74, 84-85, 88, 91-92, 95-96, 100-105, 125, 203
Silicon 183, 185, 217
Silver 133, 137, 160
Simulation 133
Software 1, 5, 17, 83, 89, 148-149, 202, 212, 217
Solar 4, 14, 41-42, 59, 80, 114, 116, 119-120, 123, 135-136, 140, 156, 160, 163, 165, 168, 171, 174-175, 182-185, 188, 196
Solid(s) 22, 28-30, 32-39, 41, 70, 140, 142-144, 163, 168
Solution(s) 8, 23, 29, 33, 54, 57, 81, 105-106, 119, 124-125, 132, 144, 148, 169, 172, 188
Sonic 21, 48
SOS 85
Sound 9, 13, 16, 21, 26, 74, 91, 95, 107, 121, 133, 174, 180
Space 1-2, 4-6, 8, 10, 13-14, 16, 21-22, 29, 37, 41, 45-47, 49-50, 62, 77, 81, 84, 88, 94-96, 105-107, 111-116, 118, 124, 126, 128-129, 132-133, 137, 144-146, 148, 150-151, 156-165, 167-168, 170-172, 175-176, 179-182, 184-185, 189, 206-209, 212-221
Space qualification 212
Space-qualified 180-181
Space shuttle 13, 22, 29, 50, 62, 126, 206, 212-213, 220
Space station 2
Spacecraft 1, 5, 8, 14, 40, 42-43, 51, 72, 74, 81, 99, 108-109, 119, 140-142, 144, 146, 149, 153-154, 157, 159-164, 166, 175, 177-179, 182-184, 188, 191, 193, 207, 209, 211, 213, 215-216, 218, 221

Speed 8, 10, 12, 14, 16, 19, 21, 26-27, 48, 56, 63, 69, 80, 87, 89, 91, 107, 114, 121, 129, 148-149
Sphere(s) 46, 72
Spin 87, 123, 126, 129-130, 132-137, 161
Spin-stabilized 133, 136
Spinner(s) 135-137
Spinning 20, 25, 129-131, 133-136, 143, 161, 184
Sputnik 47, 124
SRAM(s) 144, 146, 151-153
SSME 29
Stability 127, 161
Stabilization 5, 125-129, 132-133, 136-137, 143, 167
Stacer(s) 164
Stage(s) 29, 37-39, 136, 169-170, 203
Stanford 16, 111, 222
Star 10, 39, 41, 123, 170
Starsys 200, 203
Starsys/ORBCOMM/ELIPPSO 203
State 22, 46, 136, 140, 142-144, 147, 150, 155-156, 191, 209, 222
Static 19, 144, 146, 149-150, 155, 160, 193, 196, 216
Static-protected 194
Station(s) 2, 6, 15, 51, 57, 85, 97, 100-102, 104-106, 132, 140-141, 149, 167, 174, 202, 220
Stationary 14, 55, 58, 64, 197-198
Steel 83, 155, 163, 165
Stephan-Boltzmann 114
Storage 5, 27, 34, 39, 141-145, 173, 176, 179, 181
Strategy 78-79, 146-147
Strength(s) 41, 73-74, 77, 100-101, 132
Stress 72, 170
Structure(s) 4, 11, 14, 41, 157-158,

Index

163-164, 165, 172, 183, 191, 216
Submarines 34, 87, 199-200
Suborbital 49
Substrate(s) 80, 146, 195
Subsystems 209-210
Support(s) 38, 149, 151, 164-165, 192, 216
Surface(s) 10, 47, 52-53, 75, 100, 111, 116, 119, 124, 133, 135, 140, 157-158, 164-165, 169-170, 188, 192, 216
Survive (survival) 4, 24, 114, 143, 159, 161, 167, 172, 181, 214, 219, 222
Swedish Space Corporation 209, 221
Switch(es) 20, 79, 155, 157, 168, 171-172, 202
Synchronicity 58, 71
Synchronous 52, 54, 57-58, 63, 105
System(s) 1, 4, 11, 13-14, 27, 31, 33, 38, 42-43, 51, 57, 75, 77, 79-80, 102-104, 106, 108, 113, 125-127, 129-130, 132-133, 136-137, 139, 143, 144, 146-147-151, 149-151, 153, 156, 161-162, 168-170, 175, 187, 192, 194, 198, 202-203, 206-209, 211-212, 214, 216

T

Tank(s) 8, 11, 15-16, 22, 27, 31-32, 34, 36, 38-39, 40
Taurus 37
TDRS 4
Technique(s) 101, 125-126, 158, 164, 166
Technology (technologies) 1, 2, 4, 6, 8, 33-34, 37-38, 48, 100, 104-106, 108, 139, 143-146, 148-149, 151, 153, 169, 176, 180, 200-202, 206, 209, 211, 215-216, 222
Teflon 33
Telecommunicators 57
Telegraphy 85
Telemetry 54
Telephone(s) 4, 9, 17, 54, 56, 84, 92, 95-96, 141, 197, 203
Telescope(s) 62, 94, 100, 120, 123, 126, 156, 215
Television(s) 4, 25, 54, 84, 88, 94-96, 102, 107, 141, 188
Telstar 198
Temperature 5, 25, 27-29, 31-32, 41, 72-73, 83, 110-111, 113-116, 118-119, 135, 141, 159-161, 166, 183-184
Tensors 69
Terminal(s) 2, 4, 201, 221
Test 1, 17, 132, 150, 156, 159, 162, 165-166, 169, 172, 180, 187, 192, 196, 208, 212, 214, 216
Testing 133, 150-151, 165-166, 172, 181-182, 189, 192, 212, 214
Tetroxide 31, 39
Thermal 25, 59, 70, 116, 119, 121, 132, 150, 157, 159, 164-165, 185, 189, 212, 216
Thermal vacuum vibration 212
Thermochemistry 42
Thermochemists 29
Thermodynamic 116
Thermodynamicist(s) 24-25, 27, 29, 109-110, 113, 118
Thermodynamics 5, 25, 109
Thermometer 83
Thermostatics 109
Three-axis 75, 81, 128-129, 131-132
Thrust 14, 20, 33-34, 36-38, 40-42, 131,

237

Micro Space Craft

(Thrust, cont'd.) 136
Thruster(s) 41, 111
Titanium 39
Tolerance 144-145, 152-153
Torque-free 132-133
Torque(s) 75-80, 125-126, 130-132, 134, 136-137, 160, 216
Torquing 80, 132, 136
Toxic 29-30, 32
Toxicity 36
Toys-r-Us 173, 175, 180
Traceability 150
Track 16, 56, 64, 105, 133, 141, 143-144
Tracking 51, 84, 183-184
Trajectory (trajectories) 1, 38, 40-41, 46-47, 49, 119, 217
Transceiver 211
Transfer 33, 42, 62-63, 72, 112, 119, 121, 146, 149-150, 152-153, 211
Transistor(s) 79, 155, 157, 173, 199
Transmission 7, 54, 56-57, 85-86, 91, 94, 96, 203
Transmitter(s) 4, 56-57, 85, 87-88, 95, 99-102, 104, 119, 168, 174
Travel 7, 11-12, 16, 25, 49, 53-54, 80, 91, 107, 126

U

Ultraviolet 95, 188
Unmanned 220
Uplink 57, 102, 105, 123
Uplinking 105
Uranus 4
User 198, 207-208
Users 36, 105, 108, 201, 207, 209, 217, 221-222
USSR 51

Utah 68, 81, 209-210, 222
Utopia 217

V

Vacuum 49, 114, 132, 157-159, 161, 164-166, 170, 176, 189, 211-212, 216
Valves 20, 33-34, 38, 41
Vapor(s) 15, 29, 113, 188
Vaporware 197
Variables 15, 69-70
Variation(s) 71, 74, 76, 87, 102, 106, 118-119, 125, 136-137
Vector(s) 38, 69, 185
Vehicle(s) 8, 11-13, 14, 21, 27, 32, 37-38, 40-41, 104, 125-126, 137, 143, 159-160, 168-172, 207, 221
Velocity (velocities) 8, 13-14, 16, 20-22, 25, 28, 41, 49-50, 53, 62-63, 65-66
VHF 88, 91, 96
Vibration(s) 26-27, 54, 88, 143, 150, 159-160, 166-168, 189, 192, 196, 212, 216
Viking 221
Virginia 64, 95, 196, 212, 214
VLF 87, 91, 96
Voltage(s) 74, 77, 79, 86, 149, 177, 179, 183, 193
Volts 193
Volume 77, 141, 152-153, 208, 211
Voyager 4

Index

W

Washington 47, 123, 200, 220
Waste 75, 105, 180, 202, 209-210
Water 3, 8, 10, 13, 15-16, 22, 29, 31-32, 39, 55, 72, 75, 80, 84-85, 87, 92, 94, 107, 110, 113, 202
Watt-hours 176, 178
Watt(s) 94, 99, 100, 114, 116, 137, 154, 161, 208
Wavelength(s) 87, 91-92, 94-95, 100-101, 106, 140
Wave(s) 3, 56, 68, 87, 89, 91-92, 94-95, 101, 103, 106-107
Weather 38, 63, 69, 72, 99, 106, 126
Weber state 222
Weigh 9, 41, 168
Weighing 15, 35, 137, 177, 211, 221
Weight(s) 8, 11, 27, 33, 35, 41, 83, 94, 124-125, 143, 176, 188, 200, 207, 215
Welding 157, 161, 166, 170
Wheel(s) 7-8, 10, 20, 23, 50, 126, 129-133, 136-137, 142, 166-167, 172
Wind(s) 41-42, 77, 80, 124, 126, 128, 130, 162
Wire(s) 17, 67, 74, 76-77, 80, 84, 86, 88, 92, 107, 161-164, 191-192, 194
Wiring 86, 188

X, Y, Z

X-rays 86, 91
Xenon 33, 41
Zero gravity 32, 49, 165, 212